江苏大学英文教材基金资助出版

ASSEMBLY LANGUAGE & INTERFACING TECHNOLOGY

汇编语言与接口技术

主编
曾兰玲　韩晓茹　李　唱

副主编
【加纳】阿加加·温弗瑞德　（Adjardjah Winfred）
【加纳】库玖·派崔克　（Kudjo Patrick）

江苏大学出版社
JIANGSU UNIVERSITY PRESS
镇江

图书在版编目(CIP)数据

汇编语言与接口技术 = Assembly Language & Interfacing Technology:英文 / 曾兰玲,韩晓茹,李唱主编. —镇江:江苏大学出版社,2018.12
ISBN 978-7-5684-0855-4

Ⅰ.①汇… Ⅱ.①曾… ②韩… ③李… Ⅲ.①汇编语言－程序设计－高等学校－教材－英文②微型计算机－接口技术－高等学校－教材－英文 Ⅳ.①TP3

中国版本图书馆 CIP 数据核字(2018)第 296586 号

汇编语言与接口技术
Assembly Language & Interfacing Technology

主　　编	/曾兰玲　韩晓茹　李　唱
责任编辑	/徐　婷
出版发行	/江苏大学出版社
地　　址	/江苏省镇江市梦溪园巷 30 号(邮编:212003)
电　　话	/0511-84446464(传真)
网　　址	/http://press.ujs.edu.cn
排　　版	/镇江文苑制版印刷有限责任公司
印　　刷	/虎彩印艺股份有限公司
开　　本	/787 mm×1 092 mm　1/16
印　　张	/17.5
字　　数	/585 千字
版　　次	/2018 年 12 月第 1 版　2018 年 12 月第 1 次印刷
书　　号	/ISBN 978-7-5684-0855-4
定　　价	/50.00 元

如有印装质量问题请与本社营销部联系(电话:0511-84440882)

PREFACE

Today's digital technology, computer technology have penetrated into various fields. The ability to master and apply computer technology has become one of the criteria to measure the quality of a professional technicians. Assembly Language & Interfacing Technology is an important computer technology course in science and engineering of higher education.

Although the rapid development of computer technology, microprocessors have been from 32 to 64-bit transition, for a variety of control systems, we commonly use 8-bit, 16-bit interface technology to meet the application requirements, so 32 or even 64 microprocessor to explain the working principle of microcomputers is not suitable. This book is a 16-bit microprocessor model, about the basic principles of computer and interface technology, easy to learn and understand, its basic concepts, basic ideas and basic methods and 32-bit processor are the same.

This book is mainly for the general colleges and universities of computer science, electromechanical contror and non-electronics majors students. The preparation of this book focuses on theory with practice and engineering applications. In the preparation of teaching materials in the introduction of life , a lot of examples witch are easy to understand are used to help readers understand the basic concepts of computer. After talking about the working principle of various interface chips, the book gives a number of examples which can be directly applied, and give some different solutions to achieve the function.

The book is divided into 7 chapters. Chapter 1 introduces the basics of comicrocomprter systems; Chapter 2 introduces assembly language in graphical form; Chapter 3 introduces the assembly language programming method; Chapter 4 introduces the 16-bit processor Pins, read and write timing; Chapter 5 describes the I/O interface, the basic composition, read and write technology and data transfer; Chapter 6 introduces the concept of interrupts and functions of programming 8259A chip. Chapter 7 introduces the internal structure and programming of the programmable parallel interface chip 8255A, the serial interface chip 8251A, and the timer/counter interface chip 8253/8254.

This book is comprehensive, and self-contained, the assembly language programming part and the computer principle and interface technology are orgamied together, through appropriate choice, and is suitable both computer professional and non-computer professional students.

This book is writen by Zeng Lanling, Han Xiaoru, and Li chang; Adjardiah Winfred Kudjo Patrick as deputy editors.

The authors of this book in the process of writing referred to a lot of excellent reference materials, thanks to the author of these materials, in particular, thanks especially to Ma Weihua, Qian Xiajie and Yang Wenxian. In addition, the writing of this book has been the national teaching quality engineering network of professional comprehensive reform project and Jiangsu University teaching reform project support.

Due to the limitation of editors, there are some errors and omission, please critilize. If you encounter problems with the use of the textbook, contact the author. E-Mail: lanling@ujs.edu.cn.

Editor
August 2018

CONTENTS

CHAPTER 1 SUMMARY OF MICROCOMPUTER SYSTEM / 001

◎ 1.1　Development and application / 001
　　1.1.1　Development of electronic computer / 001
　　1.1.2　Development of microcomputer / 003
　　1.1.3　Application of microcomputer / 005
◎ 1.2　Composition of microcomputer / 006
　　1.2.1　Hardware system of microcomputer / 007
　　1.2.2　Software system of microcomputer / 009
　　1.2.3　Main performance indicator of microcomputer / 010
◎ 1.3　Number and coding of computer / 011
　　1.3.1　The conversion between the carry count system and the incoming system / 011
　　1.3.2　The representation of the number in the computer / 014
　　1.3.3　Coding in computer / 017
Exercise 1 / 019

CHAPTER 2 ASSEMBLY LANGUAGE BASICS / 020

◎ 2.1　Assembly language overview / 021
◎ 2.2　8086/8088 microprocessor programming structure / 024
　　2.2.1　8086/8088 functional structure / 024
　　2.2.2　8086/8088 memory organization / 028
　　2.2.3　8086/8088 register structure / 033
◎ 2.3　Assembly language program on the machine debugging / 038
　　2.3.1　Simple assembly language source / 039
　　2.3.2　Editing / 040
　　2.3.3　Compilation / 040
　　2.3.4　Connecting / 041
　　2.3.5　Operation and debugging / 042
◎ 2.4　Assembly language source organization / 048
　　2.4.1　Assembly language statements / 048
　　2.4.2　Assembly language source program format / 049
◎ 2.5　Operands in assembly language / 054
　　2.5.1　Constant / 054
　　2.5.2　Variables and labels / 055

  2.5.3 Expression / 062
  2.5.4 Symbol definition / 063
◎ 2.6 8086/8088 addressing mode / 065
  2.6.1 Immediate addressing / 066
  2.6.2 Register addressing / 068
  2.6.3 Direct addressing / 068
  2.6.4 Register indirect addressing / 069
  2.6.5 Register relative addressing / 070
  2.6.6 Base indexed addressing / 071
  2.6.7 Relative base indexed addressing / 071
◎ 2.7 8086/8088 instruction system / 073
  2.7.1 Data transfer class instructions / 074
  2.7.2 Arithmetic operations instructions / 083
  2.7.3 Logical and shift class instructions / 091
  2.7.4 Program control class instructions / 096
  2.7.5 Processor control instructions / 108
◎ 2.8 Basic I/O function call / 109
  2.8.1 Keypad function call (INT 21H) / 110
  2.8.2 Display function call (INT 21H) / 113
Exercise 2 / 115

CHAPTER 3 ASSEMBLY LANGUAGE PROGRAMMING / 120

◎ 3.1 Sequential programming / 120
◎ 3.2 Branch programming / 121
  3.2.1 Single branch structure / 122
  3.2.2 Double branch program structure / 123
  3.2.3 Multi-branch program structure / 125
◎ 3.3 Cyclic programming / 127
  3.3.1 Counting cycles / 128
  3.3.2 Conditional cycles / 131
  3.3.3 Multiple cycles / 133
◎ 3.4 Subroutine design / 137
Exercise 3 / 143

CHAPTER 4 16-BIT MICROPROCESSOR EXTERNAL FEATURES / 146

◎ 4.1 8086 external features / 146
  4.1.1 8086/8088 working mode / 146
  4.1.2 8086 pins / 147
◎ 4.2 8086 bus operation / 151
  4.2.1 8086 the composition of the bus cycle / 151
  4.2.2 Bus timing of 8086 / 152

◎ 4.3　8086 microprocessor subsystem / 154
　　　4.3.1　8086 subsystem in minimum mode / 155
　　　4.3.2　Maximum mode of the 8086 subsystem / 156
◎ 4.4　8088 external features / 157
◎ 4.5　80286 external features / 159
Exercise 4 / 159

CHAPTER 5　MICROCOMPUTER INPUT AND OUTPUT TECHNOLOGY / 161

◎ 5.1　I/O interface overview / 161
　　　5.1.1　I/O interface functions / 162
　　　5.1.2　Composition of interfaces / 163
　　　5.1.3　Port addressing / 164
　　　5.1.4　Classification of interfaces / 165
　　　5.1.5　Read and write technology of I/O interface / 165
◎ 5.2　I/O interface read and write technology / 166
　　　5.2.1　I/O instructions / 166
　　　5.2.2　Port composition / 167
　　　5.2.3　Address decoding in the interface / 169
　　　5.2.4　Port read and write control / 169
◎ 5.3　I/O organization / 172
　　　5.3.1　8-bit I/O organization / 172
　　　5.3.2　16-bit I/O organization / 172
◎ 5.4　Control between the interface and the host information transmission / 173
　　　5.4.1　Program control mode / 173
　　　5.4.2　Program interrupt mode / 177
　　　5.4.3　Direct Memory Access (DMA) mode / 177
　　　5.4.4　Channel mode / 178
◎ 5.5　Digital input and output / 178
　　　5.5.1　Switch output / 178
　　　5.5.2　Switch input / 181
Exercises 5 / 185

CHAPTER 6　INTERRUPTION SYSTEM FOR MICROCOMPUTERS / 187

◎ 6.1　The basic concept of interrupt system / 187
　　　6.1.1　The basic concept of interruption / 187
　　　6.1.2　Interrupt the function of the system / 188
　　　6.1.3　Interrupt handling process / 189
◎ 6.2　8086 CPU interrupt system / 192
　　　6.2.1　Classification of 8086 interrupts / 192

    6.2.2 Interrupt vector table / 194
    6.2.3 Response of 8086 to interrupt / 196
◎ 6.3 Interrupt controller 8259A / 197
    6.3.1 Pin signal of 8259A / 197
    6.3.2 The internal structure of the 8259A / 198
    6.3.3 Working process of 8259A / 200
    6.3.4 How does the 8259A work / 200
    6.3.5 Initialization of 8259A command word and initialization programming / 203
    6.3.6 Operation command word and application of 8259A / 208
◎ 6.4 Interrupt application example / 212
Exercise 6 / 216

CHAPTER 7 PROGRAMMABLE INTERFACE CHIP / 217

◎ 7.1 Programmable parallel interface chip 8255A / 217
    7.1.1 The internal structure of 8255A / 218
    7.1.2 Pin function of 8255A / 219
    7.1.3 How does 8255A work / 220
    7.1.4 Control word for 8255A / 224
    7.1.5 Application of 8255A / 225
◎ 7.2 Serial communication and serial interface / 232
    7.2.1 Mode of serial communication / 233
    7.2.2 Serial communication classification / 233
    7.2.3 Rate of serial communication / 235
    7.2.4 Serial interface standard RS-232C / 236
◎ 7.3 Programmable serial interface 8251A / 239
    7.3.1 The internal structure of 8251A / 239
    7.3.2 8251A pin function / 240
    7.3.3 8251A operate mode / 243
    7.3.4 8251A internal registers and initial programming / 244
    7.3.5 Application of 8251A / 249
◎ 7.4 Programmable timer/counting interface chip 8253/8254 / 253
    7.4.1 The internal structure of 8254 / 253
    7.4.2 8254 external pin / 255
    7.4.3 How does the 8254 work / 256
    7.4.4 Control word for 8254 / 261
    7.4.5 Application of 8254 / 263
Exercises 7 / 266

REFERENCES / 268

APPENDIX / 270

CHAPTER 1
SUMMARY OF MICROCOMPUTER SYSTEM

【Abstract】 This chapter is the basis for learning other chapters. First, this chapter introduces the development and application of microcomputer, and then focus on basic concept ,composition and main performance indicators of microcomputer system. The computer data representation and coding and basic content of logic circuit are for non-professional computer students, computer professional students can skip this part.

【Learning Goal】
- Learn the development and application of microcomputer.
- Grasp basic concept of microprocessor, microcomputer and microcomputer system.
- Learn the composition and performance indicators.
- Grasp methods of conversion from any number from base 2, base 10, or base 16 to either of the other two bases.
- Learn logic gates to diagram simple circuits.

1.1 Development and application

1.1.1 Development of electronic computer

As usually referred as computer by people, it refers to electronic digital computer. It is made of electronic devices, with abilities of computing, peforming logical judgment, processing equipments for automatic control and memory function. The world's first digital computer was inboduced in february 1946, it is an electronic numerical integrator and calculator (referred to as ENIAC). It was co-developed by physicist Mo Keli (J. Mauchly) and engineer Eckert (JPEckert) at the University of Pennsylvania.

ENIAC is the first official computer to run, but it does not have the capability of "stored programs". In June 1946, Dr. Von Neumann proposed the computer design of the "storage program" with the following:

(1) The computer hardware system is composed of five basic components, including calculator, memory, controller ,input devices and output devices.

(2) The computer uses binary representation of data and instructions.

(3) Its cores principles are "stored programs" and "program control".

The program is stored in the main memory in advance, in the case that the computer at work without the need for operator intervention, automatically remove the instructions one by one and implement.

Computer designed by this principle called Von Neumann type computer, the structure shown in Figure 1-1. The architecture proposed by Von Neumann laid the foundation of modern computer architecture theory and a milestone in the history of computer development. ① The computer architecture given by Von Neuman is minimum subset. ② The lack of any one can not become a system, a memory, a controller, an operator, an input device, and an output device. At present, the common computer in addition to Von Neumann type computer, there are Harvard structure of the computer. Harvard structure is essentially different from the Von Neumann structure, but the data and the program separately stored, with two memories, speed up the system at the same time also increased the CPU peripheral interface.

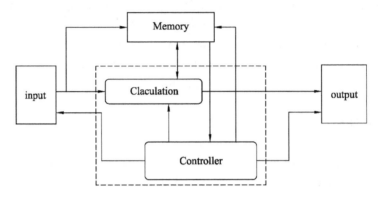

Figure 1-1　The architecture of von Neumann

From the birth of the first electronic computer to the present, computer has gone through more than half a century of development. During this period, the computer system structure is constantly changed, the application field continues to widen. Generally based on the computer used by the main physical devices, the computer development is divided into several stages, a stage called generation (Table 1-1).

Table 1-1　History of computer development

Age	First Generation 1946—1957	Second Generation 1958—1964	Third Generation 1965—1970	Forth Generation 1971—Now
Electronic Device	(electron) Tube	Transistor	Integrated circuits	Large scale integrated circuits
Memory	Delay line Magnetic core Drum tape Tape	Core Drum Tape Disk	Semiconductor memory Core Drum Tape Disk	Semiconutor memory Tape Disk CD
Processing Method	Machine language Assembly language	Monitoring program High language	Real-time processing Operating System	Real-time/time-sharing network Operating System

continued

Age	First Generation 1946—1957	Second Generation 1958—1964	Third Generation 1965—1970	Forth Generation 1971—Now
Application Field	Scientific computing	Scientific computing Dataprocessing Process control	Scientific computing System design and so on Science and technology field	All walks of life
Calculation Speed	5000 to 30000 times per second	Hundreds of thousands to million times per second	Million to several million times per second	A few million to 100 billion times per second
Typical models	ENIAC EDVAC IBM705	UNIVAC II IBM7094 CDC6600	IBM360 PDP 11 NOVA1200	ILLIAC-IV VAX 11 IBM PC

1.1.2 Development of microcomputer

The most significant event in computer development in the 1970s was the birth and rapid spread of microcomputers. A microcomputer is an organic whole that connects the microprocessor, memory, and input/output interfaces together via a bus. The microprocessor is von Neumann machine operator and controller in a chip in the organic combination, its appearance has greatly promoted the popularity of the computer.

M-E-Hoff, a young engineer at Intel Corporation in the United States, in 1969 he accepted the commission of a Japanese company to design a complete circuit for a desktop computer system. He boldly put forward a vision to put all the computers on the four chips, namely the central processor chip, the random access memory chip, the read-only memory chip and the register circuit chip. This is a 4-bit microprocessor Intel 4004, a 320-bit (40 bytes) of random access memory, a 256-byte read-only memory and a 10-bit register, they are connected through the bus, so the 4-bit microcomputer——MSC-4, 1971 was born in the United States of America. Since then, just 40 years time, the development of micro-computer has gone through seven or eight stages. People generally use the word length and the typical microprocessor chip as a sign of each stage.

The first stage (1971—1973) is the 4-bit and 8-bit low-end microprocessor era. Often referred to as the first generation, the typical products are Intel's 4004 and 8008 microprocessors and MCS-4 and MCS-8 microcomputers respectively. Basic features are mainly machine language or simple assembly language, less of instructions (more than 20 instructions), for home appliances and simple control occasions.

The second stage (1973—1978) is the 8-bit high-end microprocessor era. Commonly known as the second generation, the typical products are Intel's 8080/8085, Rockwell's 6502, Motorola's MC6800, Zilog's Z80, and a variety of 8-bit microcontroller. They are characterized by a relatively complete command system, with a typical computer architecture and interrupt, DMA and other control functions. Software in addition to assembly language, there are BASIC, FORTRAN and other high-level language and the corresponding interpreter and compiler, also appeared in the latter part of the operating system, such as CM/P is a popular operating system. This period of the more well-known computer products have 8 micro-computer TRS-80 (using Z80 microprocessor) and Apple I / II (commonly known as "Mac", using 6502 microprocessor). In addition, 8085 is more used in embedded control, 6052 and MC6800 are more used in computer game system.

The third stage (1978—1985) was the 16-bit microprocessor era. Commonly known as the third generation, its typical products are Intel's 8086/8088 and 80286, Motorola's M68000, Zilog's Z8000, and other microprocessors. It is characterized by a richer, perfect, multi-level interrupt, multiple addressing modes, segment storage mechanism, hardware multiplication and division components, and configuration of the software system.

Well-known computer products are IBM's PC (Personal Computer) in this period. The IBM PC was introduced in 1981 with 8088 CPUs. Then in 1982, the expansion of the personal computer IBM PC/XT, it has expanded the memory, and added a hard disk drive. In 1984, IBM introduced the IBM PC/AT, a 16-bit enhanced personal computer with a 80286 processor core. As IBM in the development of PC uses a technology open strategy, so that the PC swept the world.

The fourth stage (1985—1992) is the 32-bit microprocessor era. Also known as the fourth generation, its typical products are Intel's 80386/80486, Motorola's M68030/68040 and so on. It features 32-bit address lines and 32-bit data buses. 6 million instructions per second (MIPS, Million Instructions Per Second). Microcomputer function has reached or even more than super small computer, fully capable of multi-tasking, multi-user operations. At the same time, some other microprocessor manufacturers (such as AMD, TEXAS, etc.) also introduced the 80386/80486 series of chips.

The fifth stage (1993—1995) is the Pentium (Pentium) series of microprocessor era. Commonly known as the fifth generation. Typical products are Intel's Pentium family of chips and compatible with AMD's K6 series of microprocessor chips. The internal use of superscalar instruction pipeline structure, and has been independent of the instructions and data cache. With the emergence of MMX (Multi Media eXtended) microprocessors, the development of microcomputer in the network, multimedia and intelligent and so on, go to a high leve 1.

The sixth stage (1995—1999) is the use of P6 architecture family of processors and microcomputer era. Typical products are Intel's Pentium Pro, Pentium II, Pentium III and so on, the internal use of three superscalar instruction pipeline structure, the operating frequency is getting higher and higher, the bus frequency is also greatly improved. Supports multimedia extended instruction set (SIMD) MMX, SSE.

The seventh stage (2000—2007) is the NetBurst architecture Pentium 4 (Pentium 4, referred to as P4 or Pentium 4) system processor and Pentium 4 microcomputer era. As the system product performance differences, with the processor number that Pentium 4 products, typical Pentium 4 such as 5XX, 6XX, 7XX and so on. Supports SSE2, SSE3 and SIMD instructions. From 80386 to Pentium 4 early are IA-32 architecture, usually 8086/8088 and 80286 as IA-32 compatible form, no division of IA-16.

The eighth phase (2007 so far) is the use of Core architecture Core (Core) and Core Duo (Core 2) series of microprocessors. The main representatives of the product are

Core 2 Dou, Core 2 Qaurd and Core 2 Extreme and so on. The dual-core Core 2 family of microprocessors is an epoch-making new microprocessor, both 32-bit and EM64T (Extended Memory 64 Technology) technology, is a typical 32/64-bit processor. Known as the IA-32E (Incremental IA-32) architecture, which is called Intel64 because it supports 64-bit access technology.

Parallel to the fifth phase of the parallel development of pure 64-bit processors, such as Intel's Itanium, Itanium II and other processors, they use IA-64 structure. In 2008, Intel Corporation introduced the 64-bit four-core microprocessor Core i7.

1.1.3 Application of microcomputer

As the microcomputer has the advantages of small size, low price, low power consumption and high reliability, its application field is very broad. Summed up, it can be divided into two main directions. There are several application areas as follows:

1. Scientific calculation and information processing

Scientific computing refers to the use of computers to complete the scientific research and engineering technology proposed in the calculation of mathematical problems. Currently a lot of micro-machines have strong computing power, especially a number of processors constitute the system, its function can often match the mainframe, or even more than the mainframe, and the cost is low enough to make the mainframe tends to phase out. In recent years, due to the development of parallel computing technology, the parallel computing system composed of multiprocessors is very powerful in scientific computing power, and can realize various scientific computing problems which can not be solved manually.

Information processing refers to the computer on the information recording, collation, statistics, processing, use, communication and a series of activities in general. Information processing is one of the most widely used fields in computer applications. For example, the candidates in the college entrance examination work and the statistical work, railways, aircraft ticket booking system, internal cost accounting management, personnel management, salary management, financial management, contract management and banking system business management, etc., are data processing range.

2. Computer aided design, auxiliary manufacturing, auxiliary education and computer aided testing

Computer-aided design (CAD-Computer Aided Design) is the use of computer computing, logic and other functions to help people to product design and engineering technology design. Computer-aided design and auxiliary manufacturing (CAM) can be combined directly with the CAD design of the product processing out. In recent years, the industrialized countries have further developed the computerized integrated manufacturing system (CIMS-Computer Integrated Manufacturing System) at the forefront and direction of automation technology. CIMS is a large-scale computerized, automated and intelligent modern production system integrating engineering design, production process control, production and operation management. It is the future of manufacturing.

Computer-aided Education (CBE-Computer Based Education) is a computer in the field of educational applications, including computer-aided teaching (CAI), computer-aided management teaching (CMI).

Computer-aided testing (CAT) refers to the use of computers to carry out complex and extensive testing work.

3. Multimedia applications and network applications

Multimedia technology began in the 1980s. The sound, image, word processing into one, so that the computer has a computer, television, game consoles, fax machines, telephones and VCD machine integrated functions, to a machine. Computer communication is an important field of computer application developed rapidly in recent years. Through the network to achieve information sharing and remote control.

4. Process control

Process control is one of the most widely used applications for microcomputers. Currently, the manufacturing industry and daily lives, manufacturers use the computer to control automated production line, micro-machine applications in these sectors for the production capacity and the rapid improvement of product quality.

5. Embedded application direction

Embedding the microprocessor into the host application system (ie, embedded applications) to play a role is an important aspect of microcomputer applications. Microcontrollers and digital signal processors are two typical chips for this type of application. Microcontroller is known as a controller (Microcontroller), which is mainly oriented to control, in the host system as a control center; and digital signal processor is mainly for large flow of digital signal real-time processing, in the host system as a data processing center. The field of embedded applications is very extensive, from industrial production to our daily lives. For example, in the field of instrumentation, they use microprocessors to replace the traditional mechanical parts, so that it is intelligent. Again in the medical field, there are used as the core components of the microprocessor CT scanner and ultrasonic scanner, the network remote consultation. It is possible in the field of transportation, the bus train has a microprocessor everywhere, and its can achieve precise control; in life, the microprocessor-Controls, washing machine, microwave oven, refrigerator which are very popular household appliances, and microcontrollers Control of the temperature control system, automatic timekeeping, alarm have also entered the family, the phone almost become a necessity for everyone.

The current microcomputer technology is moving in two directions: one is high-performance, multi-functional direction, from this continuous achievement is the micro-machine gradually replace the expensive, superior function, the minicomputer; the other is low prices, direction of functional specificity, this development is the micro-machine in the production areas, service departments and daily life has been more and more widely used.

1.2 Composition of microcomputer

Microcomputer system consists of hardware and software, the hardware composition of the computer is called "bare metal." Bare machine is not able to run independently and handle transactions. Therefore, the hardware system must be equipped with the appropriate software system to work properly.

1.2.1 Hardware system of microcomputer

Microcomputer hardware system refers to the composition of the computer including a variety of physical devices, that is, they are the tangible physical equipment. Its basic structure and basic functions are roughly the same as large computers and small computers. However, due to the use of large-scale and ultra-large-scale integrated circuit technology and bus structure, making the micro-computer system structure has a simple, standardized and easy to expand the characteristics. The microcomputer consists of a central processor, a memory, an input/output interface and a system bus. The structure is shown in Figure 1-2. Where the central processor and memory constitute the smallest information processing unit, known as the host.

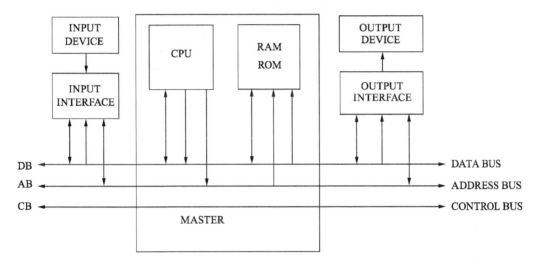

Figure 1-2 The structure of microcomputer

1. Central processing unit

The central Processing Unit (Central Processing Unit), also known as microprocessors, is the core of microcomputers, integrating the computer operator, controller and related circuits. It is mainly responsible for the implementations of instructions to achieve arithmetic and logic operations, control micro-computer components to coordinate the work.

CPU is a very large scale integrated circuit, which integrates tens of thousands of logic gate array circuit. Although the performance of various CPUs is different, is made up of the arithmetic logic components, controllers and register groups and other basic components. All the microprocessors are based on the development of these. New microprocessors are backward compatible with the characteristics of the three core components on the basis of the new design through the new components to make the CPU with storage management, multimedia and other functions.

2. Internal memory

The main task of the memory is to temporarily or permanently save the program or data resources. The memory is divided into internal memory and external memory, in which the internal memory is used to store temporary or active programs and data, and the external memory is used to permanently save the program and data.

Internal memory, also known as memory or main memory, by which the CPU can directly address the storage space, made by the semiconductor device. The memory is characterized by fast access speed. We usually use the program, such as windows operating system, typing software, game software, are generally installed in the hard disk and other external memory, but this is not its core function, they must be transferred to memory to run. We usually enter a text, or listen to a song, in fact, they are carried out in memory.

The semiconductor memory constitutes the memory and is divided into a read only memory (ROM) and a random access memory (RAM). ROM information can only be read, generally can not write, even if the machine power failure, the data will not be lost. ROM is usually used to store the computer's basic programs and data, such as BIOS ROM (Basic Input Output System). Its physical form factor is generally a double in-line (DIP) manifold. RAM information can be written and can be read. When the machine is turned off, the data stored in it is lost.

3. Input/Output interface

Input/Output interface circuit (I/O interface circuit) is a bridge between the microcomputers and external devices, and is responsible for data buffering and format conversion, coordination between the host and external equipment. The speed of data transmission differences, complete the data transfer. Different external devices through different I/O interface circuit connected with the host, such as the keyboard through the keyboard interface connected with the host, hard drive through a dedicated hard disk controller and host connection.

4. System bus

The system bus is a general between the various functional components of the public communication trunk, which is composed of wire transmission wiring harness. It is Connected to the microprocessor and memory, input and output interfaces, and it forms a complete microcomputer bus called the system bus. The system bus Retoms three different functions, such as data bus DB (Data Bus), address bus AB (Address Bus) and control bus CB (Control), system bus contains three different functions of the bus.

The data bus DB is used to transfer data information to implement data exchange between the microprocessor, the memory, and the I/O interface. The data bus is a bidirectional tri-state, which can transfer data from the CPU to other components such as memory or I/O interfaces, or to transfer data from other components to the CPU. The number of bits in the data bus is an important indicator of the microcomputer, which is usually consistent with the word length of the microprocessor.

Address bus AB is dedicated to transfer the address, because the address can only be transferred from the CPU to the external memory or I/O port, so the address bus is always one-way three-way, which is different from the data bus. The number of bits of the address bus determines the size of the memory space that the CPU can address directly. For example, the address bus of the 8-bit microcomputer is 16 bits, the maximum addressable space is 2^{16} = 64KB, and the address bus of the 16-bit microcomputer is 20 bits, Its addressable space is 2^{20} = 1MB. In general, if the address bus is n bits, the addressable space is 2^n address space (storage unit).

The control bus CB is used to transmit control signals and timing signals. Some of the control signals are sent to the memory and I/O interface circuit, such as read/write signals, interrupt response signals and so on; also some other parts sent to the CPU, such as interrupt request signal, reset signal, bus request signal and so on. Therefore, the direction of control bus transmission from the specific control signal may be, generally two-way, the control bus to the number of bits according to the actual control needs the system. In fact, the specific situation of the control bus depends mainly on the CPU.

5. External equipment

An external device refers to an input or output device, which is also a major component of the microcomputer hardware system. The input device refers to the general term of the device that transfers the information to the host, which is the means by which the host acquires the external information. The output device is the general term for the device that receives the host information, which is the means by which the host sends information to the outside. The most commonly used input devices are keyboard, mouse, scanner, light pen and so on. Commonly used output devices are monitors, printers, plotters and so on. External memory is also an external device, often both which input and output functions, known as composite devices.

1.2.2 Software system of microcomputer

In the computer system, the computer hardware system is the material basis, the computer software is the soul. Software is a computer that contains the various procedures, data and related documents required for the computer to run the computer for the effective operation and specific information processing to provide the whole process of service. It is the user to operate the computer intermediary. Software and hardware complement each other, are indispensable. In the computer development process, hardware and software systems are constantly updated and improved. Microcomputer software system is divided into system software and application software.

1. System software

System software is no special application background, its specifically explore the hardware function, test hardware components and reduce the user's dependence on the hardware and other software programs. Its main functions are to schedule, monitor and maintain the computer system; responsible for the management of computer systems in a variety of independent hardware, so that they can coordinate the work. The system software makes it possible for computer users and other software to treat the computer as a whole without having to take into account of how the underlying hardware works. System software mainly includes the following categories:

(1) Operating system. The OS reprecents the most important comparents for the computer software, it effectively manages the entire computer resources, is the computer bare metal and the application and the bridge between the users. Without it, the user can not use some kind of software or program. Typical operating systems are DOS, Windows, Linux and Netware.

(2) Service-oriented procedures. Service-oriented program refers to the software development and hardware maintenance process, auxiliary computer professionals to monitor, debug, fault diagnosis and other special procedures. Such as machine debugging, troubleshooting and diagnostic procedures, antivirus programs.

(3) Procedures for handling in various languages. It converts the various source programs written by the user in the software language into target programs that can be recognized and run on the computer, resulting in the expected results. Such as assembler, compiler and interpreter.

(4) A variety of database management systems. The database system is used to support data management and access software, which include databases, database management systems. Such as SQL Sever, Oracle, Informix, Foxpro and so on.

2. Application software

Application software refers to the programs developed and used to solve various application problems. Because the system software can not solve some specific application problems, they developed the application software. Application software is a computer-based, in the system software support, for users directly to provide users with application services, it can not only give full play to the functions of computer hardware, but also to provide users with a relaxed working environment, and can provide users with work efficiency. The Commonly used application software are as follows:

(1) Software for scientific computing.

(2) Word processing software, such as WPS 2000, Office 2003 and so on.

(3) Image processing software, such as Photoshop, 3DS MAX and so on.

(4) Various financial management, tax management, industrial control, auxiliary education software.

1.2.3 Main performance indicator of microcomputer

1. Word length

Word length refers of the computer system which can handle the CPU once the number of bits. The longer the word length indicating that the CPU can handle the higher data accuracy, the faster the processing speed, the greater the storage capacity, usually the word length is an integer multiple of the byte. The current computer word length is mainly 32-bit and 64-bit.

2. Frequency

CPU clock speed. CPU level of the frequency to a large extent determines the CPU running speed. The clock speed of the early CPU is expressed in MHz ($1M = 10^6$), and the current Pentium is clocked at GHz ($1G = 10^3 M$).

3. Operation speed

The speed of operation is the number of instructions that can be executed per second. The more commonly used measure is MIPS (Million Instructions Per Second).

4. Main capacity and access speed

The main memory capacity refers to the internal memory which can store the maximum number of bytes of data. The larger the main memory capacity, the more data that can be stored, the more programs that can be executed at the same time. The capacity of the main memory depends on the processor's addressing capability (address lines), but also by other hardware and software. For example, the Pentium 4 has 36 address lines, can access the main memory of the maximum space of 64G, but if the operating system using XP, the system only supports 4GB of memory, if you use Windows Sever 2003 can support 64GB of memory. Access speed refers to the main memory to complete a read/write the time required, the shorter the time, the faster the access speed.

CHAPTER 1　SUMMARY OF MICROCOMPUTER SYSTEM

1.3　Number and coding of computer

1.3.1　The conversion between the carry count system and the incoming system

1. Carry counting system

Carry counting system is a counting method. In the daily life, people use a variety of carry counting system. For example, hexadecimal (1 hour = 60 minutes, 1 minute = 60 seconds), decimal (1 foot = 12 inches, 1 year = 12 months). But the most familiar and most commonly used is the decimal count.

As shown in Table 1-2, the carry counting system is a method of counting using fixed digital symbols and uniform rules. Generally, K base number of the base number of K, available for the basic number of symbols are K, they are 0 ~ K-1, each digit is full of K to its high into 1, that is, "every K into 1".

The value of any number of matrices can be expressed as the sum and form of the digital sign value and the corresponding bit weight product in the binary number, which is called the polynomial of the weights. A K-ary number $(N)_K$, with the weight of the polynomial and form can be expressed as:

$$(N)_K = Dm \cdot Km + Dm-1 \cdot Km-1 + \cdots + D1 \cdot K1 + D0 \cdot K0 + D-1 \cdot K-1 + D-2 \cdot K-2 + \cdots + Dn \cdot Kn$$

(Equation 1-1)

E. g:

The decimal number $(123.4)_{10}$ can be expressed as:
$$(123.4)_{10} = 1\ 102 + 2 \cdot 101 + 3 \cdot 100 + 4 \cdot 10 - 1$$

The binary number $(1011.1)_2$ can be expressed as:
$$(1011.1)_2 = 1 \cdot 23 + 0 \cdot 22 + 1 \cdot 21 + 1 \cdot 20 + 1 \cdot 2 - 1$$

Table 1-2　Introduction to the commonly used system

System	Base	Numerical Symbols	Features	Examples
Decimal	10	0,1,2,3,4,5,6,7,8,9	"Every ten into one, by ten when ten"	$(123.4)_{10}$
Binary	2	0,1	"Every two into one, by a two"	$(1011.1)_2$
Octal	8	0,1,2,3,4,5,6,7	"Every eight into one, by one when eight"	$(35.7)_8$
Hexadecimal	16	0,1,2,3,4,5,6,7,8,9,A,B,C,D,E,F	"Every six into one, by a sixteen"	$(28A.C)_{16}$

2. Conversion between different numbers

(1) Decimal integer is converted to K-ary number

Method: in addition to K to take the remainder, the results are in reverse order.

To do this: divide the decimal number by K, get a quotient and a remainder; divide the quotient by K, and get a quotient and a remainder; continue this process until the quotient is equal to zero. The number of bits (0 to K-1) is the number of the corresponding K-number.

011

Note: the first received remainder is the lowest bit, the last received remainder is the highest bit.

[**Example 1-1**] Converts the decimal number 83 to a binary number and a hexadecimal number. The conversion process is shown in Figure 1-3.

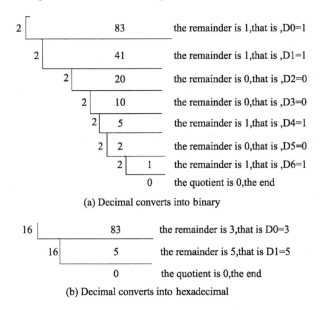

Figure 1-3 Decimal integer to K-ary integer conversion process

The final result is $(83)_{10} = (D6D5D4D3D2D1D0)_2 = (1010011)_2$
$$= (D1D0)_{16} = (53)_{16}$$

(2) Decimal decimal conversion to K decimal

Method: by K rounding, the results are in positive order.

To do this: multiply the decimal fraction by K to get a product; multiply the fractional part of the product by K and get a product; continue the process until the fractional part is 0 or the precision meets the requirement. The integer part of the product (0~K-1) is the number of the corresponding K-decimal.

Note: the first integer is the fractional high, and the resulting integer is the fractional low.

[**Example 1-2**] Converts decimal decimals 0.325 to binary decimals and hexadecimal decimals.

The conversion process is shown in Figure 1-4.

Since the integer part of the product can not be zero, it is no longer calculated. The smaller the number of bits in the fractional part, the higher the precision, the lower the bit less the accuracy.

Binary result is $(0.325)_{10} = (0.D_{-1}D_{-2}D_{-3}D_{-4}D_{-5}D_{-6}D_{-7}D_{-8})_2$
$$= (0.01010011)_2$$

Hexadecimal result is $(0.325)_{10} = (0.D_{-1}D_{-2})_{16} = (0.53)_{16}$

CHAPTER 1 SUMMARY OF MICROCOMPUTER SYSTEM

```
       0.325
    ×     2
       0.625        The integer is 0,that is , D-1 is 0
    ×     2
       1.300        The integer is 1,that is, D-2 is 1
    ×     2
       0.600        The integer is 0 ,that is,D-3 is 0
    ×     2
       1.200        The integer is 1,that is ,D-4 is 1
    ×     2
       0.400        The integer is 0,that is ,D-5 is 0
    ×     2
       0.800        The integer is 0,that is ,D-6 is 0
    ×     2
       1.600        The integer is 1,that is ,D-7 is 1
    ×     2
       1.200        The integer is 1,that is ,D-8 is 1
```
(a) Decimal to binary fraction

```
       0.325
    ×    16
       5.200        The integer is 5,that is ,D-1 is 5
    ×    16
       3.200        The integer is 3,that is ,D-2 is 3
```
(b) Decimal to hexadecimal fraction

Figure 1-4 Decimal to K fractional conversion process

(3) K conversion into a decimal number

Method: the K binary number according to the right to start.

The practice is to expand the K-ary number into the sum of the digit sign values and the corresponding bitwise product, and calculate it in decimal form. The resulting value is the decimal number.

Note: if a bit in the K-ary number is 0, the item can be omitted from the expansion.

[Example 1-3] Converts $(1010011.01010011)_2$ to decimal number.

$$(1010011.01010011)_2 = 1 \cdot 2^6 + 1 \cdot 2^4 + 1 \cdot 2^1 + 1 \cdot 2^0 + 1 \cdot 2^{-2} + 1 \cdot 2^{-4} + 1 \cdot 2^{-7} + 1 \cdot 2^{-8}$$
$$= 64 + 16 + 2 + 1 + 0.25 + 0.0625 + 0.0078125 + 0.00390625$$
$$= 83.32421875$$

[Example 1-4] Converts $(53.53)_{16}$ to a decimal number.

$$(53.53)_{16} = 5 \cdot 16^1 + 3 \cdot 16^0 + 5 \cdot 16^{-1} + 3 \cdot 16^{-2}$$
$$= 80 + 3 + 0.3125 + 0.01171875 = 83.32421875$$

(4) Binary, octal and hexadecimal conversion

In the computer, all the information is expressed in binary form, reading and writing are not convenient and prone to error. Since $2^3 = 8$, so every three binary number is equal to one octal number; $2^4 = 16$, every four binary number is equal to a hexadecimal number. So octal and hexadecimal numbers are mainly used to simplify the binary writing, and their correspondence is shown in Table 1-3. For the convenience of writing, the binary number to

"B" for the suffix, octal number to "O" for the suffix, hexadecimal to "H" for the suffix, the decimal number to "D" suffix, no suffix in decimal.

Method: the decimal point for the sector, the integer part from low to high by the 3 or 4 a group to split, the fractional part from high to low by every 3 or 4 a group of split, split into octal or ten Hexadecimal number.

Table 1-3 Correspondence between commonly used in computer

Decimal	Binary	Octal	Hexadecimal
0	0000	0	0
1	0001	1	1
2	0010	2	2
3	0011	3	3
4	0100	4	4
5	0101	5	5
6	0110	6	6
7	0111	7	7
8	1000	10	8
9	1001	11	9
10	1010	12	A
11	1011	13	B
12	1100	14	C
13	1101	15	D
14	1110	16	E
15	1111	17	F
16	10000	20	10

Note: after segmenting less than 3 or 4 bits of the integer part of the high fill 0, the fractional part of the low fill 0.

[Example 1-5] Converts $(1010011.01010011)_2$ to octal and hexadecimal numbers.
$$1010011.01010011B = 001\ 010\ 011.010\ 100\ 110B = 123.246O$$
$$1010011.01010011B = 0101\ 0011.0101\ 0011B = 53.53H$$

Since the octal can be easily converted to a 3-digit binary number, when converting a decimal number to a binary number, the decimal number can be easy converted to an octal number and a hexadecimal number.

1.3.2 The representation of the number in the computer

1. Unsigned number

The so-called unsigned number is a set of positive number and zero. When storing a positive number or 0, all bits are used to store the number of digits, regardless of its symbol, and the unsigned number is named.

When an unsigned number is represented by N-bit binary, the smallest number is 0 and the maximum number is 2^N-1. 8-bit binary unsigned number of the range of 0 ~ 255, 16-bit binary unsigned number of the range of 0 ~ 65535.

If an unsigned number needs to increase its number of bits, it is necessary to add a number of 0s on its left side, called zero expansion. For example, when storing 8-bit unsigned number 0110 1001B with 16 bits, add 8 0s to the left of the number, and the result is 0000 0000 0110 1001B (insert space is for reading and distinguishing, and there is no need for writing).

2. Signed number

In fact, in more cases, the computer handles the number with symbols. In the computer, the positive and negative sign of the signed number is represented by the most significant bit of the data: 1 means that the symbol is negative and 0 indicates that the symbol is positive. For the convenience of the narrative, the first introduction of two nouns: the number of machines and the true value. The number of expressions in the machine, that is, the number of machines called the number of machines, the number itself called the true value. There are three commonly used machines: original code, complement code and anti code.

(1) Original code

The original code is the highest bit as a sign bit, the positive number, the bit 0, the negative number, the bit 1. While the numerical part holds the original form (sometimes need to add a few bits in the upper part). The result is the original representation of the number.

[**Example 1-6**] X = +1001010B, Y = -1001010B, Z = -1110B(= -0001110B). When the original code is 8-bits, X, Y and Z of the original code are:

$[X]_{original}$ =01001010B; $[X]_{original}$ =11001010B; $[X]_{original}$ =10001110B.

The original code that there have three main features: one is intuitive, and the true value of the conversion is very convenient; the second is to multiply, in addition to the operation is convenient; three is plus, minus the operation more trouble.

(2) Complement code

Positive number, complement code with the original code. For example, X = +0101001B, $[X]_{complement} = [X]_{original}$ =00101001B. The complement of the negative number is equal to the original code in addition to the sign bit outside the bit "inversion" (1 change 0, 0 change 1), the last bit plus 1. For example, Y = -0001100B, $[Y]_{original}$ =10001100B, $[Y]_{complement}$ = 11110011B +1 =11110100B. It can be seen from the method of complement code, for the complement, the sign bit and the original code symbol bit, also said the true value of the symbol. The purpose of the introduction of the complement is to facilitate the addition and subtraction of the signed number.

(3) Inverse code

The positive number is the same as the original code and the same as the complement. On the negative number, the inverse code is equal to the original code in addition to the sign bit, according to bit reversed (the last bit without 1). Generally, the inverse code is used as the intermediate process of complement, that is, $[x]_{complement} = [x]_{inverse} +1$.

The machine number coding described above is mainly used for assembly language programming. In high-level languages, numbers can be signed, but the compiler eventually represents it as a machine number.

Table 1-4 lists the 8-bit binary code that is part of the value of the original code, inverse code and complement. From the table can be seen 0 of the original code and inverse code have two kinds of representation, can not express −128. And the complement of the +0 and −0 representation of the method is unique, which can be more than the number −128.

Table 1-4 8-bit machine number of the original code, inverse code and complement representation

True Value (Decimal)	True Value (Binary)	Original Code	Inverse Code	Complement Code
+127	+111 111	0 111 111	0 111 1111	0 111 1111
+1	+000 0001	0 000 0001	0 000 0001	0 000 0001
+0	+000 0000	0 000 0000	0 000 0000	0 000 0000
−0	−000 0000	1 000 0000	1 111 1111	0 000 0000
−1	−000 0001	1 000 0001	1 111 1110	1 111 1111
−2	−000 0010	1 000 0010	1 111 1101	1 111 1110
−127	−1111111	1 111 1111	1 000 0000	1 000 0001
−128	−1 000 0000	Can't be expressed	Can't be expressed	1 000 0000

3. The operation of the binary number

(1) The operation of the unsigned number

The unsigned arithmetic is actually an operation between positive numbers, and all digits are all valid data bits, with no sign bits. Table 1-5 lists the operations of the unsigned number of bitwise operations.

Table 1-5 Binary rules

Operation	Operation Rules
Additon	0 +0 =0;0 +1 =1;1 +0 =1;1 +1 =0(HAVE CARRY)
Subtraction Rules	0 −0 =0;1 −1 =0;1 −0 =1;0 −1 =1(Have a borrow)
Logic ADD	0 ∧0 =0;0 ∧1 =0;1 ∧0 =0;1 ∧1 =1
Logic OR	0 ∨0 =0;0 ∨1 =1;1 ∨0 =1;1 ∨1 =1
Logic XOR	0 ⊕0 =0;0 ⊕1 =1;1 ⊕0 =1;1 ⊕1 =0
Logic NOT	$\bar{0}$(−) =1; $\bar{1}$(−) =0

(2) The addition and subtraction of the signed number in the computer with its complement form to participate in the operation, we first discuss the complement of the rules of operation.

① The rules of the complement code

The rules for the complement code are as follows:

A. $[X+Y]_{complement} = [X]_{complement} + [Y]_{complement}$, that is, the sum of the two numbers is equal to the complement of the complement.

B. $[X-Y]_{complement} = [X]_{complement} + [-Y]_{complement}$, that is, the difference between the two numbers are equal to the complement of the complement of the complement and the complement of the sum of the complement.

C. $[[X]_{complement}]_{complement} = [X]_{original}$, that is, by the method of complement code, $[x]$ complement the complement again, the result is equal to $[x]$ original.

D. $[[X]_{complement}]_{complement} = [-X]_{complement}$, fill operation is to include the sign bit plus 1 plus the operation.

Where b and d show that as long as the complement of the $-Y$ complement $[-Y]$, you can turn subtraction into addition. This means that in the CPU as long as there is an adder which can be added and subtraction operations, it can simplify the CPU circuit.

[Example 1-7] Let $X = +100$, $Y = +83$, and find $[X-Y]_{complement}$.

Solution: first seek $[X]_{complement}$ and $[-Y]_{complement}$.

$$[X]_{complement} = [+100]_{complement} = [+1100100B]_{complement} = 01100100B$$
$$[-Y]_{complement} = [[Y]_{complement}]_{complement} = [[01010011B]_{complement}]_{complement}$$
$$= [01010011B]_{complement} = 10101101B$$

$$\begin{array}{r} [X]_{complement} \\ +[-Y]_{complement} \\ \hline [X-Y]_{complement} \end{array} \quad \begin{array}{r} 01100100B \\ +10101101B \\ \hline 1\ 00010001B = 11H = 16+1 = 17 \end{array}$$

Carry is naturally lost

[Example 1-8] Let $X = +100$, $Y = +83$, find $[Y-X]_{complement}$.

Solution: first seek $[-X]_{complement}$ and $[Y]_{complement}$。

$$[-X]_{complement} = [[X]_{complement}]_{complement} = [[100]_{complement}]_{complement} =$$
$$[01100100B]_{complement} = 10011100B$$
$$[Y]_{complement} = [83]_{complement} = [+1010011B]_{complement} = 01010011B$$

$$\begin{array}{r} [Y]_{complement} \\ +[-X]_{complement} \\ \hline [Y-X]_{complement} \end{array} \quad \begin{array}{r} 01010011B \\ +10011100B \\ \hline 11101111B = -0010001B = -(16+1) = -17 \end{array}$$

Note: the difference between the complement code and the complement of the operation: positive complement and the original code is the same, the negative complement is equal to its original code in addition to the bit outside the bit "inversion" (1 change 0,0 variable 1), the last bit plus 1. The complement operation does not distinguish between positive and negative numbers, and all bits, including the sign bits, are negated.

1.3.3 Coding in computer

Modern computers not only deal with numerical data, but also deal with a large number of non-numerical data, such as English letters, punctuation marks, special symbols and Chinese characters. No matter what information must be binary coded to be recognized and processed by the CPU. Several common binary coding methods are discussed below.

1. ASCII code

The ASCII code, the American Standard Cord for Information Interchange, is the most commonly used code in a computer. It uses 128 bits and 12 characters and symbols. See Appendix A for details.

ASCII encoding has the following characteristics:

(1) Each character is represented by 8 bits (ie, one byte), where the most significant bit

is "0", and when the parity is required, the most significant bit is used as the parity bit.

(2) ASCII encoding a total of 128 characters, they mainly include: 0~9 ASCII code for the 30H~39H; 26 English capital letters A~Z ASCII code 41H~5AH; 26 English lowercase ASCII Code for the 61H~7AH; 32 control characters, mainly used for communication in the communication control or computer equipment, functional control, encoding 0~31.

2. BCD code

BCD (Binary Coded Decimal) code is a decimal number expressed in binary code. It retains the right of the decimal number, that is, every ten; and the number is 0 and 1 combinced to represent. Decimal representation of binary numbers requires at least 4 bits of binary code. 4-bit binary coding has 16 combinations, and decimal only 10 symbols, choose 10 symbols to represent the decimal 0~9 has a variety of feasible programs. The following only describes the most commonly used BCD code.

8421 BCD code is the most basic and most commonly used BCD code, it is similar to the four natural binary code, you have the weight of 8,4,2,1, so called the right BCD code. And four natural binary code are different, it only uses the four binary code in the first 10 groups of code, that is, with 0000~1001 on behalf of its corresponding decimal number, the remaining six groups of code do not. Table 1-6 gives the correspondence between the BCD code and the decimal and hexadecimal.

Table 1-6 BCD code and other binary control table

Binary Code	BCD Code	Decimal	Hexadecimal	Binary Code	BCD Code	Decimal	Hexadecimal
0000	0000	0	0	1000	1000	8	8
0001	0001	1	1	1001	1001	9	9
0010	0010	2	2	1010	Illegal	No	A
0011	0011	3	3	1011	Illegal	No	B
0100	0100	4	4	1100	Illegal	No	C
0101	0101	5	5	1101	Illegal	No	D
0110	0110	6	6	1110	Illegal	No	E
0111	0111	7	7	1111	Illegal	No	F

BCD code has compressed BCD code and non-compressed BCD code two encoding methods. Each bit of the compressed BCD code is represented by a 4-bit binary, and one byte represents a two-digit decimal number. For example, 10010110B that decimal number 96D; non-compressed BCD code a byte can store a decimal number, where the high 4-bit content is not specified (also part of the book requirements for 0, both can), low 4-bit binary indicates the decimal number. The ASCII code 37H (00110111B) of the numeric character '7' is the uncompressed BCD code of the number 7 (the contents of the upper 4 bits are not specified).

CHAPTER 1   SUMMARY OF MICROCOMPUTER SYSTEM

Exercise 1

1.1   Which five components constitute von Neumann type computer ?

1.2   What are microprocessors, microcomputers and microcomputers? What are their compositions?

1.3   What is the system bus? According to the different types of signals, system bus can be divided into what three types of system bus? What are their characteristics?

1.4   What are the indicators that measure the performance of microcomputer systems?

1.5   Convert decimal numbers $(123.025)_{10}$ and $(96.12)_{10}$ into binary and hexadecimal numbers.

1.6   Converts hexadecimal numbers $(3E.7)_{16}$ and $(A4.B)_{16}$ to decimal numbers.

1.7   Find 8-bit unsigned number 10110101B and 01011101B corresponding to the decimal number?

1.8   Find respectively 78 and -78 8-bit original code, complement code and anti code.

1.9   Set $[X]_{complement} = 11001010B$, $[Y]_{complement} = 01001010B$, find their true value.

1.10   Let $X = +37$, $Y = -15$, find $[X-Y]_{complement}$.

CHAPTER 2
ASSEMBLY LANGUAGE BASICS

【Abstract】 This chapter first introduces the 8086/8088 programming structure, and elaborated 8086/8088 CPU memory organization, the internal functions and the use of each register. In order to make the reader master the assembly language programming method as soon as possible, this chapter begins with a simple assembly source program, step by step and a general view of explaining the assembly language programming on the machine debugging method, this is to help readers have an intuitive understanding of the language programming and how to debug the program and view results. Again, this chapter focuses on the common assembler directive, the addressing mode supported by the 8086/8088 system, and the common instruction functions and applications of the 8086/8088 instruction system.

Directive is the basis of assembly language programming, only skilled in the 8086/8088 instructions of the writing format, function and precautions, programmers can skiled operate in the preparation of assembly language procedures process readers, can use the debugger (such as DEBUG) to observe the effect of instruction execution, can be more intuitive and in-depth understanding of the instruction function.

【Learning Goal】
- Master the 8086/8088 programming structure: understand the memory organization, logical address and physical address conversion, familiar with the commonly used register name and role, and the meaning of the flag register flag.
- Familiar with the assembly language format and source code format.
- Master the assembly language program development and debugging methods.
- Master constant expression methods, variable definitions, variable attributes and their applications.
- Master and apply 8086/8088 addressing mode and instruction system commonly used instructions.
- Understand the transfer range of the target address and addressing mode.
- Familiar with the correct use of the keyboard and the display of the commonly used DOS function call.

CHAPTER 2 ASSEMBLY LANGUAGE BASICS

2.1 Assembly language overview

Although it is difficult to fully understand the characteristics of assembly language before programming in assembly language, it is useful to understand that the characteristics of assembly language are useful for learning assembly language programming. The application of computer has penetrated into all areas of social life, and has become "language" that people communicate with the computer. There are three main stages of machine language, assembly language and high language, and is moving in the direction of "natural language". Assembly language is a low-level language that takes full advantage of the hardware characteristics of the computer, and it has a very close relationship with the computer architecture. While high-level languages can achieve the functionality that most machine languages can achieve. The assembly language is often used to improve the productivity of computer software and hardware control systems, as well as for high-level language debugging, providing efficient code for computer systems.

1. Machine language

The machine language is a set of machine instructions that are encoded in binary, and is a machine-oriented language consisting of 0 and 1 codes. It is the only language that the CPU can directly identify and execute. The machine language description program is called the target program.

Machine instructions and CPU are closely related. Usually, different types of CPU correspond to different machine instructions. Different types of CPU instruction set are often a big different. But the same series of CPU instruction set often has good backward compatibility. For example, the Intel 80386 instruction set contains the 8086 instruction set.

The following set of instruction code is the use of Intel 8086 machine instructions to achieve the addition of two blocks, including three machine instructions, expressed in hexadecimal form as follows:

```
A01020
02061120
A21220
```

The function of the block is difficult to see directly. These three machine instructions add the number in the offset address 2010H unit in the memory to the number in the offset address 2011H unit, and the result of the operation is sent to the offset address 2012H unit.

The machine language can not be described in familiar ways, the machine language programming is very difficult, easy to error, once it is wrong, it is difficult to find, not easy to read and maintain. But it is the language program that can help computer to directly identify and execute. Because the machine language is compiled by the program it is not easy to understand, and finchans communication process. So, only in the early or when the machine language to write the program, and now, almost no people use the machine language to write the program.

2. Assembly language

In order to overcome the shortcomings of the machine language, people use a symbol that is easy to remember and can describe the instruction function to represent the opcode of the instruction, which is called the instruction mnemonic. The mnemonic is generally an abbreviation for the English vocabulary or vocabulary of the instruction function. Also use symbols to represent operands, such as CPU registers, memory cell addresses, and so on.

Using assembly language, the above two numbers of added program fragments can be expressed as follows:

```
MOV    AL,[2010H]
ADD    AL,[2011H]
MOV    [2012H],AL
```

Obviously, the assembler format instruction is much easier to grasp than the binary code machine instructions. Assembly language is a computer language that uses the mnemonics to represent instruction functions. The instructions in mnemonics are called assembly instructions. Programs written in assembly instructions are called assembly language source programs, or simply assembler source programs, assembly language programs. Assembly language programs are easier to understand, debug, and maintain than machine language programs.

The only language that the computer can recognize directly is the machine language consisting of 0 and 1 codes, so the source program written in assembly language must be translated into the target program represented by the machine language before being executed by the CPU. The process of translating assembly language source programs into target programs is called assembly. The program that completes the assembly task is an assembler or assembler. The assembly process is shown in Figure 2-1.

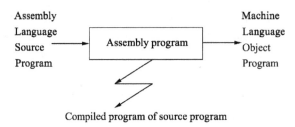

Figure 2-1 Diagram of assembler

3. High-level language

Although the assembly language with symbol replaces machine instructions, to a certain extent, simplifies the preparation of the program, reading and debugging. It still requires that programmers are familiar with the internal structure of the computer, the preparation of assembly language program or a certain degree of difficulty, not conducive to the computer promotion and application.

A high-level language is a programming language similar to natural language and mathematical description language. It's oriented to the process or object-oriented, from the specific machine. Easy to write, read and debug, and portability is good. But the high-level language source program must be compiled to be translated into a program which can identify, the implementation of the target program, usually can not produce a valid machine language code, resulting in its large memory space, the program runs slower.

4. Assembly language features

Assembly language using mnemonic form, compared with the machine language, assembly language is easy to understand than the memory. The assembly language program is also easy to write, read and debug. In particular, the assembly language has the following main features:

(1) Assembly language and machine are closely related

Assembler format instructions are symbolic representations of machine instructions, one-to-one correspondence with machine instructions, and assembly language is still a low-level language for machines. So the assembly language source program compared with the high-level language source, its versatility and portability is much worse. But through the assembly language can be the most direct and most effective control of the machine, which is often difficult to do most of the high-level language.

(2) Assembly language program is highly efficient

The assembly instructions that make up the assembly language are the "avatars" of a corresponding machine instruction that directly controls the hardware of the computer and takes full advantage of many of the features of the machine hardware system. So the target program is highly efficient, reflected in the time is the implementation of fast; reflected in the space is that the target program is short, less memory footprint. In the "space-time" efficiency, using the same algorithm under the premise of high-level language program tends to be much less.

(3) Writing assembly language source code is cumbersome

Writing assembly language source programs are much more cumbersome than writing high-level language source programs. Assembly language is a machine-oriented language, high-level language is a process-oriented or object-oriented, object language.

The operation that can be done by each assembler format instruction that is symbolized by the machine instruction which is extremely limited. Programmers use assembly language to create procedures, need to consider almost all the details of the problem, including registers, storage units and addresses. Such as the effect of the instruction execution on the flag, the location of the stack settings, and so on. When writing programs in high-level languages, programmers do not encounter these trivial but important issues.

(4) Assembly language program debugging is relatively difficult

Debug assembly language programs are often more difficult than debugging high-level language programs. The limited functionality of assembler format instructions and the fact that programmers should pay attention to too many details are the objective causes of this difficulty.

5. Assembly language applications

Based on the characteristics of assembly language, its applications are:

(1) The occasion that the software execution time or storage capacity of the higher requirements, for example, the core of the operating system, intelligent instrumentation control system, real-time control system.

(2) The occasions that the need to improve the performance of large-scale software. The subroutines (processes), which are frequently executed in large software, are usually written in assembly language and then connected with other programs.

(3) Software and hardware are closely related, the occasion of the software needs to

directly and effectively control the hardware. Such as device drivers, initialization blocks for I/O interface circuits, and so on.

(4) Occasions that there is no suitable high-level language or only use assembly language. For example, when developing the latest processor program, there is no compiler that supports the new instruction.

(5) Other, such as the underlying software system, encryption and decryption software, analysis and prevention of computer virus software.

2.2 8086/8088 microprocessor programming structure

The programming structure, refers to the programmer and the user from the perspective of the structure, and the internal structure of the chip and the actual layout of the difference. Microprocessor is a large-scale integrated circuit technology made of semiconductor chips, being the core components of the computer system. The internal components of the computer includes: the controller, the operator and the register group. The Intel 8086/8088 microprocessor is a representative high-performance 16-bit microprocessor in the Intel family of microcontrollers with a 16-bit data path and pipelined architecture that allows parallel operation on prefetch instructions when the bus is idle, the execution and implements instructions.

In the application-oriented principles, this section focuses on the 8086/8088 microprocessor programming structure. The following will be the functional structure, memory organization, register group and other content expansion.

2.2.1 8086/8088 functional structure

From the functional point of view, Intel 8086/8088 microprocessor can be divided into two parts namely, the execution unit EU (Execution Unit) and the bus interface unit BIU (Bus Interface Unit), as shown in Figure 2-2. EU is not connected to the external bus, it is responsible for the implementation of the instruction decoding. BIU is responsible for reading instructions and read/write data from memory or external devices, that is to complete all bus operations. These two units are in parallel to a certain degree of working state, improving the speed of the microprocessor to execute the instruction. This is the simplest instruction pipeline technology.

CHAPTER 2 ASSEMBLY LANGUAGE BASICS

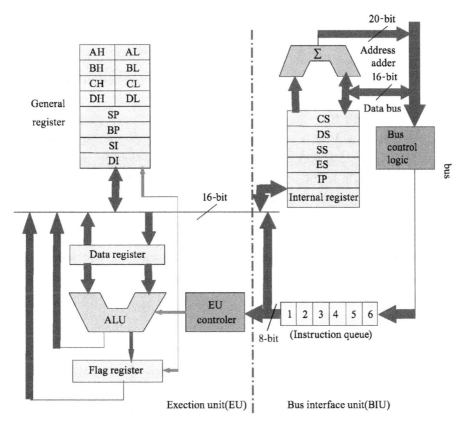

Figure 2-2　8086 programming structure

1. Execution unit (EU)

The eexecution unit (EU) is responsible for decoding, executing, including arithmetic, logic operations, control, et al. The instruction execution unit (EU) includes an arithmetic logic operation unit (ALU), a 16-bit flag transmitter (FR), eight 16-bit registers, a data register and an EU controller.

(1) Arithmetic logic operation unit (ALU)

It is a 16-bit operator for arithmetic or logic operations. The result of the operation is sent to the general register via the internal bus or to the internal register of the BIU, waiting to be written to the memory. The features of the result after operation of ALU (with or without carry, overflow, etc.) are placed in the flag register. The input data register is used to temporarily store the operands for the operation.

(2) Flag register (Flags)

It is also known as the program status word register (PSW), used to store the features of the results and control signs after ALU operation The flag register is 16 bits, and 9 bits are actually used in the 8086/8088 CPU.

(3) General register group

It contains eight 16-bit registers that hold operands or addresses. Each register has its own specific purpose.

(4) EU controller

The EU controller is responsible for fetching instructions from the instruction queue of the BIU and decrypting the instruction. The corresponding command is sent to the EU internal parts according to the contents of the instruction to complete the function specified by each instruction. It is equivalent to the controller in the computer.

2. Bus interface unit BIU

The bus interface unit BIU is responsible for all access operations of the 8086/8088 for memory and I/O devices. Specifically, it is responsible for fetching instructions from the memory unit and sending it to the instruction queue buffer for temporary storage; reading the operand from the memory unit or peripheral port or transferring the execution result of the instruction to the specified memory unit or set the port; Form a physical address (PA) according to the effective address (EA).

The bus interface unit BIU includes four segment registers, one instruction pointer register, internal register, first-in first-out instruction queue, bus control logic, and address adder for calculating 20-bit physical address.

(1) 4-bit, 16-bit segment registers

They are used to store the different logical segment of the segment start address respectively.

(2) 16-bit instruction pointer register (Instruction Pointer, IP)

IP is used to store the next offset address EA (also called an active address) to be executed next, and IP is only combined with CS to form a physical address pointing to the instruction storage unit. During the execution of the program, the contents of the IP are automatically incremental modified. When the EU executes the branch instruction and the instruction is invoked, the IP address is the destination address.

(3) Instruction queue

The role of the instruction queue is to store the instruction code fetched from the memory by the BIU. When the EU is executing instructions and does not need to occupy the bus, the BIU automatically performs prefetch instructions. 8086 instruction queue is 6 bytes, and store 6 bytes of instruction code according to the order. The queue register is operated as "first in, first out".

(4) Bus control logic

The bus control logic connects the internal bus of the 8086 microprocessor to the external pins and is the only way for 8086 microprocessors to exchange data with memory or I/O ports. The external pins of the 8086 are described in Chapter 4.

(5) Address adder

8086 microprocessor output 20-bit physical address can be directly addressed 1MB storage space, but the CPU internal registers and internal bus are 16-bit registers. The 20-bit physical address is formed by a dedicated address adder. The address adder adds 4-bit to the left of the segment register contents (the start address of the segment) to the 16-bit offset address. For example, when the instruction is fetched, the address adder adds the contents of the code segment register (CS) 4 bits to the left and 16 bits after the IP value to form a 20-bit physical address, and sends the address bus fetch instruction.

3. BIU is coordinated with EU operations

8086 introduced instruction pipeline technology, with instruction-level parallelism. BIU and EU coordinate the following principles:

(1) The execution unit EU does not directly connect outside, but instead obtains the instruction from the instruction queue of the bus interface unit BIU and executes it. Whenever an instruction is enforced in the instruction queue, the EU immediately begins execution.

(2) Whenever there are 2 empty bytes in the instruction queue, the BIU automatically searches for the idle bus cycle for prefetch instructions until it is filled. The order of the fetch is the order in which the instructions appear in the program.

(3) Whenever the EU is ready to execute an instruction, it fetches the code from the front of the instruction queue of the BIU component and then executes the instruction in several clock cycles. In the process of executing the instruction, the memory or I/O port must be accessed. The EU will request the BIU to enter the bus cycle to form the effective address EA to the BIU. The BIU address adds forms the physical address PA from the memory or I/O port fetches operand to EU to complete access to memory or I/O ports. If the BIU is taking an instruction word into the instruction queue, the BIU will first complete the fetch bus cycle and then respond to the EU's request. If the BIU is fetching an instruction word into the instruction queue, the BIU will immediately respond to the EU bus request. Access to bus request.

(4) Whenever the EU executes a transfer, call or return instruction, the BIU clears the instruction queue buffer and fetches the fetch instruction from the new destination address into the instruction queue. The EU can continue to execute the instruction and implement the execution of the program. At this time, the parallel operation of EU and BIU is obviously affected, but the probability of calling instruction is not very high as long as it is transferred. The working mode of cooperation between EU and BIU will greatly improve the working efficiency of CPU.

(5) When the command queue is full, and the EU does not have a bus access request, BIU will enter the idle state.

In summary, BIU and EU are a parallel. BIU and EU are independent of each other and cooperate with each other, in the EU's implementation of the instructions at the same time, BIU can prefetch the next or a few instructions. Therefore, in general, the CPU executes the next command stored in the command queue immediately ofter executing one command, thereby reducing the CPU to take the instruction to wait for the time to improve the CPU utilization, speed up the machine speed of operation.

4. 8088 programming structure

From the perspective of the internal structure of the microprocessor, the 8088 is similar to the 8086. The difference is only in the following two aspects:

(1) The data pins of the 8088 and the external exchange data are 8 bits, the data bus width between the bus control circuit and the special register set is also 8 bits, and the EU internal bus is still 16 bits, so the 8088 is also called "quasi-16 Microprocessor."

(2) The length of instruction queue in 8088 BIU is only 4 bytes. As long as there is one free byte in the queue, BIU will automatically access the memory prefetch instruction to fill up the instruction queue.

2.2.2　8086/8088 memory organization

1. Memory address space and data storage format

8086/8088 has 20 address bus, so the addressable address space capacity is 220 (1M). The memory is organized in bytes (8 bits), with each byte corresponding to a unique 20-bit address (ie, the actual address of the storage unit, also known as the physical address or absolute address). The address of the memory cell is continuous, the address is numbered from 0, and the order is incremented by 1, which is an unsigned binary integer, usually written in hexadecimal form. The entire address space ranges from 0 to -1 (00000H to FFFFFH), as shown in Figure 2-3.

Hexadecimal address	Binary address	Memory
0 0 0 0	0000 0000 0000 0000	
0 0 0 1	0000 0000 0000 0001	
0 0 0 2	0000 0000 0000 0002	
0 0 0 3	0000 0000 0000 0003	
.	.	
.	.	
.	.	
F F F E	1111 1111 1111 1110	
F F F F	1111 1111 1111 1111	

Figure 2-3　The address of the memory

The information stored in a storage unit is called the contents of the storage unit. As shown in Figure 2-4, the address of 00002H, 00003H storage unit content were 12H, 34H, which can be expressed as:

(00002H) = 12H, or [00002H] = 12H
(00003H) = 34H, or [00003H] = 34H

In the 8086/8088 system, multi-byte data in the memory of a number of consecutive storage units, low byte into the low address, high byte into the high address (ie, "high and low" storage principle, also known as Small byte or small end.); Expression, with its low byte address that multi-byte data occupied by the address space. For example, in Figure 2-4, the contents of the address 00002H word unit are expressed as: (00002H) = 3412H. Address 00002H The contents of the double word unit are expressed as: (0002H) = 78563412H.

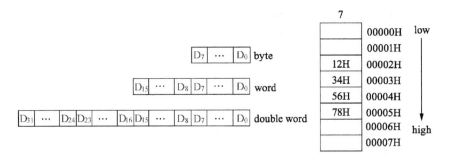

Figure 2-4　8086 Data storage format

Special note: a storage unit address can be seen as the byte unit, word unit, as well as double word unit, therefore, the use is determined by the specific circumstances.

The numbering method of the memory cells is incremented from right to left from 0, the LSB is D_0, the most significant bit (MSB) corresponds to byte, word, double word D_7, D_{15}, D_{31}, as shown in Figure 2-4.

In the 8086 storage system allows word to start from any address. If a word is stored from an even address, it is called a rule word or an alignment word. If a word is stored from the odd address, it is called an irregular word or a non-aligned word. Access to rule words can be done in one bus cycle, with two bus cycles for irregular word access.

2. The segmentation of the memory

8086/8088 system address bus has 20 lines, and its addressable memory address space is 1MB. However, 8086/8088 microprocessor internal registers and internal bus are all 16-bit, which can only address up to 64KB of space. How can I use 16-bit registers or internal bus to address 1MB of memory space? In order to solve this problem, 8086/8088 system introduces the concept of storage space logical segmentation, that is, the entire 1MB of storage space is divided into several segments, each segment has a independent logical unit called logical segment can be independently addressed by memory. 8086/8088 memory logic segmentation follows the two rules:

(1) The starting address of each logical segment (referred to as the segment address/segment base address/segment header) must be modulo 16 address (xxxx0H), that is, the logical address of the first storage unit 20 bits of the lower 4 bits must be 0. In this way, the lower 4 bits that are fixed to 0000B are omitted, leaving only the upper 16 bits, and the segment address can be represented by 16-bit binary numbers. In the 8086/8088 CPU, there are four 16-bit segment registers dedicated to store different logical segment addresses.

(2) The maximum length of each segment does not exceed 64KB (the maximum addressing space represented by the 16-bit register is 64KB). Aside the logical segmentation, the address of a memory cell in the segment can be represented by an offset relative to the starting unit of the segment. This offset is called the intra-segment offset address, also known as the effective address (EA). Since each segment does not exceed 64KB of storage space, the offset address can also be represented by a 16-bit binary number.

Based on the characteristics of the above logical segmentation, we can uniquely locate the storage unit for any of the storage units as long as we know the segment address and its offset relative to the first memory cell of the segment. That is, using the "segment address: segment address offset" we can local the specified storage unit, this address is called the logical address. The logical address is 8086/8088 CPU internal and the address form is used in the program design. The segment address and the offset address in the segment are represented by an unsigned 16-bit binary number.

8086/8088 system accesses any one of the storage unit, the address bus must be transmitted on the 20-bit physical address. Obviously, the physical address of the access memory cell should be derived from the logical address translation. So how does the 16-bit segment base address and the 16-bit segment offset address change to a 20-bit physical address to access the memory?

It can be seen from the characteristics of the logical segmentation that the lower four bits of the 16-bit segment address are four "0", that is, the lower 4 bits of the fixed 0000B which are originally omitted when the segment address is restored, unit 20-bit physical address, and then add the 16-bit offset address to get the corresponding access to the storage unit 20-bit physical address. The conversion relationship is shown in Figure 2-5, where the segment address is shifted to the left by 4 bits, plus the offset address in the segment, which is equivalent to the following address calculation:

Physical address = segment address ×16 + offset address in segment

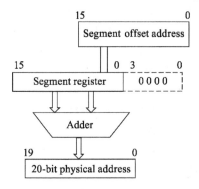

Figure 2-5 The formation of physical addresses

When the CPU accesses the memory, it must complete the above address calculation, which is done by the 20-bit address adder in BIU.

The segments of the storage space are not unique, and the logical segments can be connected to each other, or they can be completely separated or overlapped (partially overlapping or completely overlapping). Any of the storage units can be uniquely contained in a logical segment or in two or more overlapping logical segments. Thus, the same physical address can correspond to multiple logical addresses, which can be obtained by combining different segment addresses and offset addresses. In other words, for any storage unit, the physical address is unique, but the logical address is not unique, there can be multiple. As shown in Figure 2-6, where the physical address of the storage unit is 23162H, the segment address values of the two overlapping segments are 2002H and 2012H respectively, and the offset addresses in the corresponding segments are 3142H and 3042H respectively.

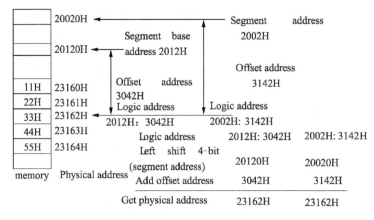

Figure 2-6 Logic and physical address

3. The relationship between the segmentation of information and the segment register

The logical segmentation of the memory facilitates the storage of information by feature segmentation. In the 8086/8088 system, there are four logical segments, each of which has its own purpose. In order to save the segment address of the corresponding logical segment, the 8086/8088 is designed with four 16-bit registers: the code segment register CS, the data segment register DS, the stack segment register SS, and the additional segment register ES. Each segment register is used to store the segment address of a logical segment, so that four segments can be used at the same time, as shown in Figure 2-7.

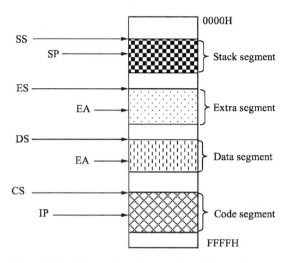

Figure 2-7 Segment and segment register reference

(1) Code Segment: the code segment stores the program sequence of instructions.

The code segment register CS stores the segment address of the current code segment. The instruction pointer register IP stores the offset address of the instruction to be executed in the code segment. The CPU uses CS: IP to fetch the instruction to be executed. That is, the contents of CS left 4 plus the contents of the IP is the next instruction to be executed in the memory of the physical address. For example, if the offset address of an instruction in the code segment is 0100H, ie (IP) =0100H, the current code segment register (CS) =2000H, the physical address of the instruction in the main memory is PA =(CS) 4 bits +(IP) =20000H + 0100H =20100H.

(2) Data Segment: holds the data (variable) used currently by the running program.

The data segment register DS stores the segment address of the current data segment. The offset address (EA) of the memory operand is calculated as in the memory addressing mode of the instruction (detailed in section 2.6 of this book). CPU using DS: EA read and write memory data segment data, that is, the contents of the DS left 4 plus the calculated EA, you can get the corresponding data stored in the physical address (PA).

(3) Stack Segment: it is the main memory area where the stack is located. It is the temporary data storage area (stack area) required for program execution, and adopts the "last in first out" mode of operation.

The stack register SS stores the segment address of the current stack segment. One side of the stack section is fixed, whiles the other side is movable the fixed side is called the bottom of the stack, and the movable side is called the top of the stack. Once the stack segment is defined, the system will automatically indicate the top position of the stack (that is, the offset address at the top of the SP storage stack) with the SP as a pointer. When a data pops up or push on the stack, only the PUSH and POP instructions use SS: SP to handle the top of the stack. The physical address of the stack at this time should be:

PA = (SS) left shift 4 bits + EA

(4) Extra Segment: is an additional data segment that is also used for data storage.

The additional segment register ES stores the segment address of the additional segment. The string operation instruction takes the additional segment as the storage area of its destination operand.

Each segment register is used to determine the start address of a segment, each segment has its own purpose. The segment register referenced by the instruction and the stack operation is defined as CS and SS respectively. It is immutable. The segment register specified by the destination operand in the string operation is also defined as the ES. However, when accessing the general memory operand, the segment register may not be the default DS; when the BP register is involved in the addressing mode, the segment register may not be the default SS, as long as the segment register is specified directly with the segment override prefix, you can change both of these defaults. 8086/8088 design has the following four segments beyond the prefix:

CS: the code segment goes beyond the use of the code segment of the data.

SS: the stack segment goes beyond the data using the stack segment.

DS: data segment transcends the data of the data segment.

ES: additional segment beyond, using additional segment data.

Table 2-1 lists the section register reference. Please note, that the situation allows the paragraph beyond.

Table 2-1 Segment register reference specification

Methods involved to access memory	Default segment register	Superable segment address	Offset address
Fetch instruction	CS	NONE	IP
Stack operation	SS	NONE	SP
General data access (except for the following)	DS	CS,ES,SS	Valid address
Source data sting	DS	CS,ES,SS	SI
Destination data string	ES	NONE	DI
BP is used as a pointer register	SS	CS,DS,ES	Valid address

In general, when the program is less, the amount of data is not large, the code segment, data segment, stack segment and additional segments can be set in the same paragraph, that is included within 64KB. When the program and data volume is larger, more than 64KB, you can define multiple code segments, data segments, additional segments and stack segments. At this time in CS, DS, SS and ES is stored in the current use of the logical segment address, the use of these sections can be modified by the contents of the register to access the other segments to expand the scale of the program. If necessary, you can point to other segments by adding a segment override prefix to the instruction.

2.2.3 8086/8088 register structure

As shown in Figure 2-2, there are 14 16-bit registers in the microprocessor. These registers can be divided into three main groups based on its functions: general register group, segment register group, and control register group. The register structure is shown in Figure 2-8.

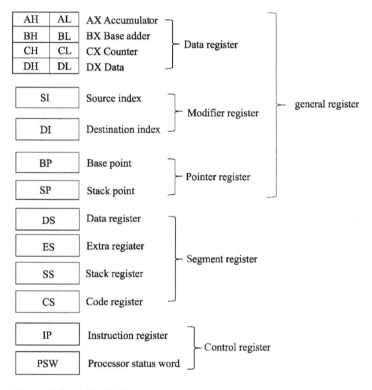

Figure 2-8 8086/8088 register structure 1(general register group)

1. General register group

The general purpose registers are divided into three groups: a data register, an address pointer register, and an index register.

(1) Data register

The 8086/8088 has four 16-bit data registers:AX, BX, CX, DX. They can be used as a 16-bit register, store data or address, or as two 8-bit registers. The lower 8 bits are called AL, BL, CL, DL, and the upper 8 bits are called AH, BH, CH, DH. As the 8-bit register can only use the data stored, can not store the address. The duality of these registers makes the 8086/8088 microprocessor handle the word can also handle byte data.

(2) Index register

The 8086/8088 has two 16-bit index registers: SI, DI. They are usually used in conjunction with the DS to provide a segmented offset address or offset address component for accessing the current data segment. In the string operation instruction, SI specifies the offset address used to store the source operand, called the source index register. The DI specifies the offset address used to store the destination operand, called the destination index register. Since the string operation instruction specifies that the source operand (source string) if must be in the current data DS, the destination operand (ie, the destination string) must be in the additional data segment ES, so that in the string operation, SI, DI must be associated with DS, combined with ES. When SI, DI are not used as a register, they can also be used as a general data register to store operands or operations. Note that they can only be used as a 16-bit register and can not be used as an 8-bit register.

(3) Address pointer register

The 8086/8088 has two 16-bit address pointer registers: SP, BP. They are usually used to store the offset address of the stack operand. SP and BP are often used in conjunction with SS to facilitate access to the current stack segment. Where SP is called the stack pointer register, the SP is always stored in the current stack unit of the offset address, the stack operation PUSH and POP instruction are derived by the SP offset address. BP is the base pointer register, BP is stored by default in the stack of a storage unit offset address or offset address of the component. When the operand is in the stack, BP can be used as the base register to read and write the data in the stack segment using random access.

The BP register can also be used as a general data register to hold operands or operations. Note that it can only be used as a 16-bit register and can not be used as an 8-bit register. The SP register can only be used as a stack pointer register, but not as a data register.

The above eight 16-bit general-purpose registers are generic in general, and they also have their own specific usage, and some instructions also implicitly use these registers. For example, the string operation instruction and the shift instruction specify that the CX register (other registers) can not be used as the count register, so that it is not necessary to give the CX register number in the instruction, shorten the instruction length, and simplify the writing of the instruction. It is commonly said that this use is "implicit addressing". Programmers must follow these instructions when programming. Table 2-2 shows the special purpose and implicit feature of the general registers in the 8086/8088. Assembly language programmers must pay full attention to these uses in order to use these general registers correctly and rationally.

Table 2-2 Specific uses of general purpose registers

Register name	Specific usage	Implied feature
AX, AL	· Used as an accumulator in multiplication and division instructions · Used as a data register in I/O instructions	Implicit addressing Explicit addressing
AH	Used as a destination register in LAHF	Implicit addressing
AL	In the BCD code operation instructions for the accumulator	Implicit addressing
BX	· Used as an address register or a base register in memory addressing · In the XLAT instruction for the base address register	Explicit addressing Implicit addressing

continued

Register name	Specific usage	Implied feature
CX	In the loop instruction and string operation instructions as a loop counter, each time a cycle, CX content minus 1	Implicit addressing
CL	Used as a counter for the number of shifts in the shift and cyclic shift instructions	Explicit addressing
DX	· Used as an address register when the I/O instruction is indirectly addressed · In the multiplication and division instructions as auxiliary counters (when the product or dividend is 32 bits when the high 16 bits)	Explicit addressing Implicit addressing
BP	Used as the base register for accessing the stack segment	Explicit addressing
SP	Used as a stack pointer in the stack operation	Implicit addressing
SI	· The source index register is used as a string manipulation instruction · Used as an address register or index register in memory addressing	Implicit addressing Explicit addressing
DI	· Used as a destination index register in string manipulation instructions · Used as an address register or index register in memory addressing	Implicit addressing Explicit addressing

2. Segment register group

In the program design, the address code of the access memory consists of the segment address and the offset address in the segment. The segment register is used to store the segment address. The bus interface unit (BIU) is equipped with four 16-bit segment registers, the code segment register (CS), the data segment register (DS), the stack segment register (SS) and the additional segment register (ES). The CPU can access four different segments of memory in four segment registers. The use of the segment register is described in Section 2.2.2 for the segmentation of the information and the segment register.

3. Control register group

(1) Instruction pointer register (Instruction Pointer, IP)

The instruction pointer register (IP) is similar to the function of the program counter PC in the traditional CPU and is used to store the offset address of the next instruction to be executed in the current code segment. During program execution, the contents of the IP are automatically modified by the BIU so that it always points to the next instruction address to be executed, so it is an important register used to control the execution order of the instructions. Its content program can not be directly accessed, but when the implementation of the program control class instructions, the contents can be implicitly modified into the target address.

(2) Flag register (FLAGS)

The flag register is also called the program status word (PSW) register. 8086/8088 CPU has a 16-bit flag register, used to store the results of the operation and the machine working status, the actual use is only 9, the specific format shown in Figure 2-9. Flags can be divided into two categories according to the function: a class called the state flag, is used to indicate the characteristics of the operation results, the instructions are automatically established after, a total of six: CF, PF, AF, ZF, SF and OF. These features may affect the operation of subsequent instructions. The other is called the control flag, which is used to control the operation of the CPU or working state, a total of three: DF, IF and TF. The control flag is artificially set, and the instruction system has instructions that are specifically designed to set or clear the control flag. Each control flag has a control effect on a particular operation of the CPU.

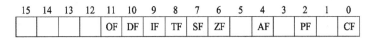

Figure 2-9 8086/8088 flag register format

◇ Status flag

① (Zero Flag, ZF)—zero mark. If the result of this operation is 0, then ZF = 1, otherwise ZF = 0.

② (Sign Flag, SF)—sign sign. This flag shows whether the symbolic result is positive or negative. For the signed number, the highest bit of the symbol, if the operation of the results of the highest bit is 1, then SF = 1, otherwise SF = 0.

③ (Parity Flag, PF)—parity sign. This flag is PF = 1 when the number of "1" in the lowest byte of the operation result is 0 or even. Note that the PF flag only shoos that the number of "1" bits in the lowest 8 bits of the operation result is even or odd, even for 16-bit word operations.

④ (Auxiliary to Flag, AF)—Auxiliary carry flag. AF = 1 when there is a addition/subtraction in the lower 4 bits (ie, D3 bit to D4 bit) when performing 8-bit (byte) or low 8-bit operation of 16-bit (word) AF = 0. AF flag is mainly for the BCD code decimal arithmetic instructions to determine whether to do decimal adjustment, the user generally do not care.

⑤ (Carry Flag, CF)—carry flag. This flag is set to "1", ie, CF = 1 if the result of this arithmetic operation results in the carry (addition) or borrow (subtraction) of the most significant bit. If the most significant bit of the result of the addition is not carried, or the subtraction result is the highest bit without borrow, then CF = 0.

⑥ (Overflow Flag, OF)—overflow flag. Occurs when the result of the operation overflows, OF = 1, otherwise OF = 0.

The so-called overflow, occurs, when the operation of the signed number, the byte operation results beyond the range of -128 ~ +127, or word operation results beyond the range of -32768 ~ +32767, known as overflow. Because the result of the operation has exceeded the range of values that can be represented by the target unit, which will result in the loss of a valid number and an erroneous result.

It is noted that the overflow flag OF and the carry flag CF are two different meanings. The carry flag indicates whether the result of the unsigned operation is out of range and the result is still correct. The overflow flag indicates whether the signed result is out of range and the result is incorrect.

When the processor operates on two operands, the result is obtained according to the unsigned number, and the carry flag CF is set accordingly. At the same time, the overflow flag

OF is set according to whether or not the signed number is exceeded. Which logo should be used is decided by the programmer. In other words, if the operand the participate in the operation that is unsigned, it should be concerned about the carry. That is a results in an signed number, should pay attention to whether the overflow.

There is a simple rule to determine whether the result of the operation overflows: an overflow occurs only when the same number of symbols (including the number of different symbols) is subtracted, and the result of the operation is opposite to the original data symbol, The result is obviously not correct. In other cases, no spill occurs.

For example, the hexadecimal numbers 53H and 46H are added.

$$
\begin{array}{r}
0101\ 0011\quad (53\text{H}) \\
+\ 0100\ 0110\quad (46\text{H}) \\
\hline
1001\ 1001\quad (99\text{H})
\end{array}
$$

The result is negative, it is clear that the operation has an overflow, that is, OF =1; since the most significant bit of the result is 1, SF =1; the result of the operation is not 0, Z =0; the result of the lower 8 bits is an even number, so PF =1; the most significant bit of the result does not produce a carry forward, so CF =0, the process of D3 to D4 Low 4 bits to the upper 4 bits) generated carry, so AF =1.

OF and CF discriminant analysis: In this case, if it is considered unsigned, 83 +70 =153, still within the range of 8-bit unsigned number (0 ~ 255), no carry, ie CF = 0. If it is considered a signed number, 83 +70 =153, has exceeded the 8-bit signed number of table range (− 128 ~ +127), resulting in overflow, that is OF = 1. Also, the result 99H is expressed as a complement of the true value of -103, and it is clear that the result of the operation is incorrect because the correct result of the two positive numbers can not be negative.

Also, add the hexadecimal numbers 0AAH and 7CH.

$$
\begin{array}{r}
1010\ 1010\quad (0\text{AAH}) \\
+\ 0111\ 1100\quad (7\text{CH}) \\
\hline
10010\ 0110\quad (26\text{H})
\end{array}
$$
The highest bit of carry 1 is lost.

It is not possible to overflow, that is, OF =0; since the most significant bit of the result is 0, SF =0; the result of the operation is not 0, so ZF =0; the lower 8 bits of the result, the number of 1 in the number is odd, so PF =0; the most significant bit of the result is generated forward, so CF =1, the process of D3 bit to the D4 bit generated carry, so AF =1.

OF and CF difference analysis: In this case, if it is considered unsigned, that is 170 +124 = 294, has exceeded the 8-bit unsigned number of table number range (0 ~255), generating a carry, that is, CF =1. If it is considered a signed number, that is −86 +124 =38, still in the 8-bit signed number of table number range (−128 ~ +127) within, no overflow, that is, OF =0.

◇ Control signs

① (Interrupt Enable Flag, IF)—Interrupt enable flag. IF =1, it indicates that the CPU is allowed to respond to an external maskable interrupt request; if IF =0, the CPU is disabled from responding to an external maskable interrupt request. The IF flag can be set to "1" with the STI instruction, and the CLI instruction clears the IF flag by "0".

② (Direction Flag, DF)—direction flag. It controls the direction of change of the string pointer. If DF = 0, the string operation instruction causes the address pointer to be incremented automatically, that is, the string operation is carried out from the low address to the high address. If DF = 1, it indicates that the address pointer is automatically decremented, that is, the high address to the low address. The DF flag can be set to "1" with the STD instruction, and the DF flag can be cleared to "0" with the CLD instruction.

③ (Trap Flag, TF)—single step flag. TF = 0, indicates that the CPU normally executes the program. If TF = 1, it means that the CPU into the single-step work mode, that is, the CPU every time the implementation of a command to automatically generate a number 1 internal interrupt (this internal interrupt is called single-step interrupt, so TF called single step flag), So that the CPU go to the implementation of a single step interrupt service routine. The use of single-step interrupt can be a one-step instruction to debug the program, this one by one instruction debugging method is single-step debugging. The user can use this function to check the execution of each instruction. This is useful during program debugging.

2.3 Assembly language program on the machine debugging

Assembly language learning is inseparable from machine experiment. Assembly language program on the machine operation includes four steps: editing, assembly, connection, commissioning operation (Figure 2-10).

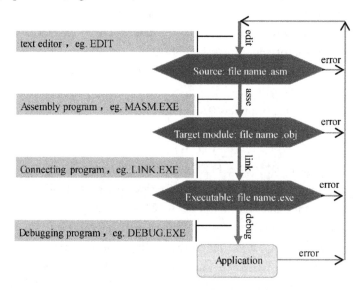

Figure 2-10 assembly language program on the machine debugging process

Common assemblers are MASM (Marco Assembler) and TASM (Turbo Assembler). Connectors are LINK and TLINK. Debuggers are DEBUG and TD (Turbo Debugger). MASM, LINK and DEBUG are produced by Microsoft Corporation. TASM, TLINK and TD are produced by Borland. Taking Microsoft's assembly toolkit as an example, this book introduces the debugging process of assembly language program on the computer.

Note: In this section, we will focus on the process of debugging procedures and methods, hence we do not need to explore the examples of the program code and the meaning of the instructions. Relevant content is described in detail later.

2.3.1 Simple assembly language source

Here is a complete simple assembly language source.

[**Example 2-1**] The string "Hello, Assembly!" is displayed on the screen.

```
; Source: ex201.asm
; Function: Display the string "Hello, Assembly!" On the screen
DSEG SEGMENT                        ; data section begins.
    STRING DB 0DH, 0AH, 'Hello, Assembly!', 0DH, 0AH, '$'
DSEG ENDS                           ; data segment ends
CSEG SEGMENT                        ; code section begins
    ASSUME CS: CSEG, DS: DSEG       ; Sets the relationship between the segment
                                    ;   register and the logical segment
START: MOV AX, DSEG
       MOV DS, AX                   ; Sets the data segment address.
       MOV DX, OFFSET STRING        ; Sets the entry parameters for DOS function
                                    ;   calls
       MOV AH, 9                    ; Set the function number for the DOS
                                    ;   function call
       INT 21H                      ; DOS function call, display string
       MOV AX, 4C00H                ; Sets the function number and entry
                                    ;   parameters for the DOS function call
       INT 21H                      ; DOS function call, return to the DOS
                                    ;   operating system
CSEG ENDS                           ; code segment ends
    END START                       ; assembly end
```

8086/8088 system, the assembly language based on the logical segment, according to the concept of segment to organize the code and data. Normally, the data variable is defined in the data segment and the program is written in the code segment. The logical segment definition uses a pair of pseudo instructions SEGMENT and ENDS, which do not generate machine code, it is used to indicate the start and end of a logical segment. The segment is named by the programmer. Example 2-1 defines two logical segments with segment names DSEG and CSEG, respectively.

The ASSUME pseudo instruct statement on line 5 tells the assembler that the CS register corresponds to the CSEG segment and the DS register corresponds to the DSEG segment from now on. DSEG is used as a data segment and CSEG is used as a code segment. So the program code is placed in the CSEG segment.

The last line END START directive tells the assembler to compile the source code at the end of the source code. That is, the end of the compilation. END is the reserved word, START is the same as the starting point in line 6, and the START symbol is the entry address. The code after the END directive will not be compiled into the object code. The label is also named by the programmer.

Here we combine Example 2-1 Introduction assembly language program on the machine debugging process.

2.3.2 Editing

The editing phase of the task is: input assembly language source; the source program to modify.

Any text editing software can be used to enter and modify the assembly language source program, such as the command line mode full-screen text editor EDIT, other high-level language programming tools in the editing environment, Windows Notepad (Notepad), write Board (Writer), Word and so on. Be sure to use the plain text format to save the assembler source file, otherwise it can not be assembled. Assembly language source files should generally be ASM for the extension, which can simplify the follow-up steps in the operation of the order.

Select "Start—Programs—Accessories—Command Prompt" and click the Start DOS command window. In this window, we can switch between the full screen and the window by pressing the Alt + Enter shortcut at the same time to facilitate the operation. In the command line mode, the specific operation shown in Figure 2-11. After the command is entered, the carriage return takes effect.

Figure 2-11 Switch to the masm folder

We can also use Notepad and other text editing environment to complete the source code entry, the correct input Example 2-1 program code, save the source file ex201. asm. Can enter the next compilation link.

2.3.3 Compilation

The task of the compilation phase is to translate the source program into a target module file (.OBJ) consisting of machine code.

If there is no syntax error in the source program, MASM will automatically generate a target module file (ex201.obj); otherwise MASM will give the corresponding error message. At this time should be based on the error message, re-edit the source code, and then compilation.

Assembly ex201.asm source specific operation: input masm ex201.asm ↙. As shown in Figure 2-12.

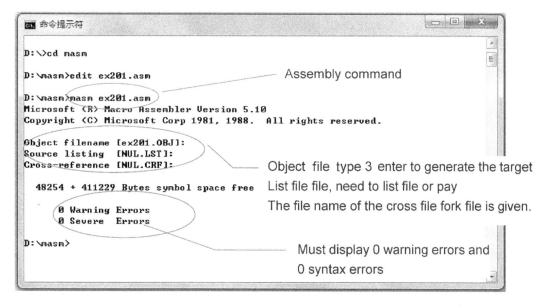

Figure 2-12 Assemble the ex201. asm file interface

If use the semicolon " ; " end command (such as: masm ex201. asm;), the assembler no longer prompts the input module file name, list file name, etc. , directly using the default file name. By default, the source file name is the same as the source file, and the extension is the extension of the corresponding type file, such as the target module file (. obj) and the list file (. lst).

2.3.4 Connecting

The connection phase connects one or more object files and library files into a complete executable (. EXE, . COM file).

To connect the ex201. obj target module file to an executable file: Enter link ex201. obj. As shown in Figure 2-13.

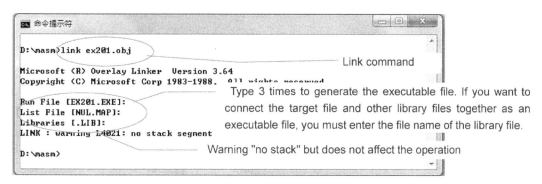

Figure 2-13 Connect the ex201. obj file interface

If there is no error, LINK will generate an executable file (ex201. exe); otherwise it will prompt the corresponding error message. In this case, you need to re-modify the source file according to the error message and then compile and link until the executable file is generated.

2.3.5 Operation and debugging

The assembly connection generated by the executable program just depends on the input file name, carriage return to run, as shown in Figure 2-14.

Figure 2-14 Run the ex201 file interface

The operating system loads the file into the main memory, and start running, Example 2-1 run the effect as shown in Figure 2-14. If there is a run error, you can start troubleshooting from the source, you can also use the debugger to help find the error.

Based on Figure 2-14, we find that, the results of the implementation of the assembly language program has the following two different methods, that is applicable to different assembly language procedures.

① Directly in the DOS command, enter the executable file name of the assembler to observe the implementation of the results, such as Example 2-1 assembler. This method is applicable to assembly language programs that run results directly on the screen.

② Use the DEBUG command to observe the results of the program, this method is applicable to the implementation of the assembler. The contents of the various registers or memory units of the machine must be carefully observed.

The main commands of DEBUG are listed in Table 2-3. Note the use of common debugging commands.

Table 2-3 DEBUG main order

Command format	Function Description
A [address]	Assembly
C [range] address	Memory area comparison
D [range]	Displays the contents of the memory unit
E address [byte value table]	Modify the contents of the memory unit
F range byte value table	Fill the memory area
G [=Start address] [Breakpoint address table]	Breakpoint execution
H value value	Hexadecimal addition and subtraction
I port address	Enter from port
L [address (Drive Number Sector Number Number of sectors)]	Read from disk

continued

Command format	Function Description
M range address	Memory area transfer
N File identifier [File identifier…]	Specify the file
O port value	Output to port
P [=address] [value]	Execution process
Q	Exit DEBUG
R [register name]	Display and modify contents of register
S range byte value table	Search in memory area
T [=address] [value]	Tracking execution
U [range]	Disassembly
W [address [drive number sector number number of sectors]]	Write to disk

1. DEBUG command instructions

(1) The DEBUG command instructions accept and display the number are expressed in hexadecimal, and do not need to give the suffix letter H.

(2) Command is a letter, the command parameters vary with the command.

(3) Commands and parameters are not case sensitive.

(4) Separators (spaces, tabs, commas, etc.) which are only between two adjacent hexadecimal numbers is necessary, between the command and parameters can not be separated.

(5) When the prompt appears, you can type the DEBUG command, only after pressing the Enter key, the command will start execution.

(6) If DEBUG checks out a command syntax error, use "^ Error" to indicate the wrong location.

(7) Use Ctrl + Break key or Ctrl + C key to interrupt the execution of a command, return DEBUG prompt.

(8) If a command produces a considerable number of output lines, in order to see the display on the screen, you can press Ctrl +S key to pause the display.

2. DEBUG command parameter description

In addition to exiting the command Q, other DEBUG commands can have parameters.

(1) Address, address parameters usually represent a memory area (or buffer) start address, which consists of segment address and offset address. The segment address can be represented by a segment register, or it can be represented by a 4-digit hexadecimal number. The offset address is represented by a 4-digit hexadecimal number. There must be a colon between the segment address and the offset address. In addition to the A, G, L, T, U, and W commands implicitly using the value of the CS register, other commands implicitly use the value of the DS register.

(2) Range, the scope for the specified memory area (buffer), by two ways: the first is to use the starting address and the end of the address that the end address can not have a segment address. The second is represented by the start address and length. The length must be guided

by the letter L. The maximum range is 64K, that is, 0 to 0FFFFH.

E. g:

    CS: 100 110

    CS: 100 L10

The following procedure uses the DEBUG. EXE debugger to observe program 2-1 program execution. Specific steps are as follows.

① Enter the DEBUG state

When the program 2-1 program compiles, and the connection is successful, that is generated executable file ex201. exe, enter "DEBUG ex201. exe ↙", loading ex201. exe, enter the DEBUG state. Note that you always have to add the file suffix (. exe), otherwise it will be an error. As shown in Figure 2-15.

② Disassemble the U command

Use disassembly the U command to display the contents of the memory unit as machine instructions, in the form of mnemonics. Use disassembly the U command to view the disassembly of the program ex201. exe and determine the end address of the program. As shown in Figure 2-15.

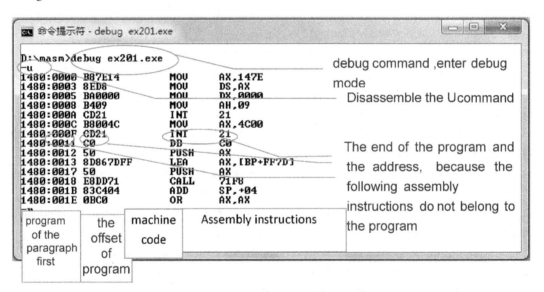

Figure 2-15　DEBUG U command

The U commands with no arguments start at the beginning of the current CS: IP, or immediately after the last disassembly end address, with a length of 32 bytes which are displayed on the screen. The U command display is divided into three parts: the program occupies the storage address, machine code and assembly instructions. U commands with parameters can be disassembled from the address specified by the parameter.

③ R command

After the previous step is performed in the memory, you get command position and can begin to prepare, but before the implementation, you need to look at the original register and memory-related unit content, which can be compared after the implementation of the relevant unit Whether the content is loaded or changed correctly.

Use the R command to display the contents of the 8086/8088 registers and the next instruction to be executed, as shown in Figure 2-16. The contents of each register are related to the actual use of the memory.

DEBUG uses a method of display the flag status symbol to reflect the flag value. The status of each flag is indicated by two letters respectively. The symbols represents the eight flag states as listed in Table 2-4.

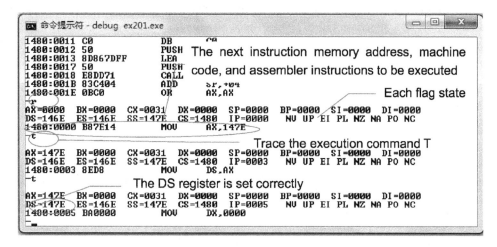

Figure 2-16　DEBUG R, T command

Table 2-4　Symbol representation of the flag state in DEBUG

Logo name	Oversize OF	Direction DF	Interuption IF	Symbol SF	Zero ZF	Auxiliary carry AF	Parity PF	Carry CF
Set state	OV	DN	EI	NG	ZR	AC	PE	CY
Reset state	NV	UP	DI	PL	NZ	NA	PO	NC

The R command does not only displays the contents of each register, but if also modifies the contents of the general-purpose registers and segment registers, as well as the instruction pointer IP and flag registers.

④ Trace the execute command T

After looking at the register value, first, execute the two T commands to set the DS register value correctly. As shown in Figure 2-16, the two T commands perform the following two instructions, respectively:

```
MOV AX, DSEG
MOV DS, AX         ; Sets the data segment address
```

Visible from above, the instructions after assembly can directly with the data segment address 147EH instead of DSEG. The MOV instruction is a data transfer instruction, which is the most used instruction in the program. The assembler instruction format is:

```
MOV dst, src       ; dst ← (src), passing the source operand src to the
                     destination operand dest.
```

Therefore, from the first two instructions after the implementation, we can see (DS) = (AX) =147EH, the correct completion of the data segment address settings.

Use the trace execution command T to track the execution of one or more instructions. The T command specifies the start address and the address argument is numbered, such as the absence of a segment address in the address parameter, then the CS is the segment address. It must be noted that, the start address must be an executable instruction. If there is no start address, then the trace execution starts from CS:IP. If the number of trace execution instructions is not specified, then follow the implementation of instruction. The T command will follow the DOS function caller. Please note that under normal circumstances the DOS function caller and BIOS program is not entered.

⑤ Display the memory unit command D

Set the DS register value correctly, first, look at the initial content of the data segment, that is, the definition of the variable storage allocation. Use the command D to view the contents of the data segment (segment address must be based on the actual load address, in the previous step has been correctly loaded), as shown in Figure 2-17. The display is the left of the memory unit logical address bar. The middle part of the byte value (hexadecimal). The right is the byte value as ASCII code corresponding to the symbol, for non-ASCII code, or non-display symbols, with a point or space. We can see the case of 2-1 definition of the string stored in memory from the figure.

If the DS command is not executed before the DS command is loaded, the 4-bit hexadecimal value should be given directly to the parameter segment address of the D command. DS can not be used for the segment address.

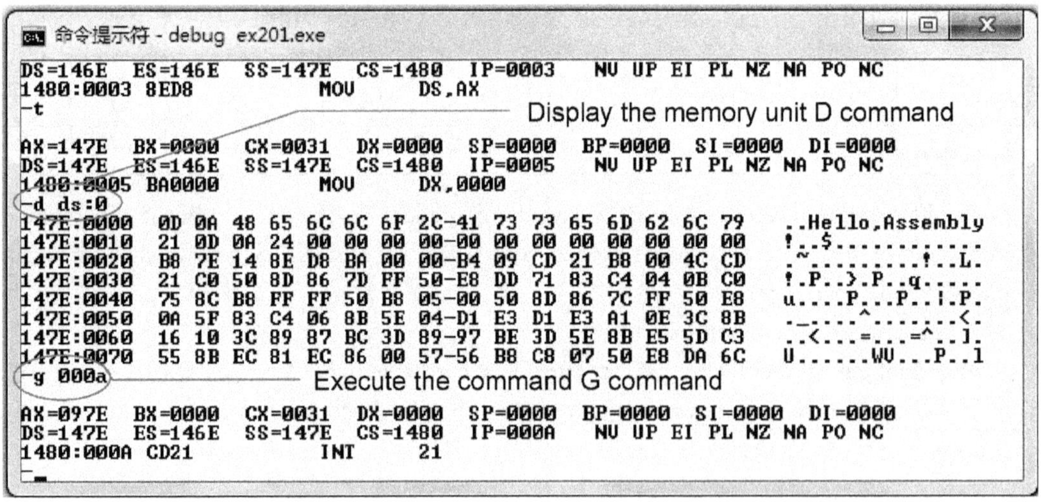

Figure 2-17 DEBUG D,G command

⑥ Execute command G

After observing the data segment, we use the breakpoint execution command G to run multiple instructions consecutively. As shown in Figure 2-17, this step we specify the breakpoint for the offset address 000AH instruction "INT 21H" (instruction in memory in the storage location we are in debug mode step 2 U command view, as shown in Figure 2-15).

The G command controls execution of the instruction from the current CS: IP until the breakpoint position 000AH is paused. Of course, you can also specify other breakpoints.

Use the execute command G to set the breakpoint to execute the debugger. The G command that does not specify the start address is executed from the current CS: IP until the end of the breakpoint or program is terminated.

⑦ Execute the process command P

The instruction next to be expellented is the instruction "INT 21H" for the offset address 000AH unit. Note that this time it is best not to use the T command, you can use the P command.

As shown in Figure 2-18, the instruction "INT 21H" is executed with the P command. At this time, only the last two instructions are not executed. Finally, execute the T command and the P command in turn to end the program execution. Note the information "Hello, Assembly!" Displayed by the first P command after executing the DOS function call. The second P command calls the DOS 4CH function to terminate the program execution, so the DEBUG displays the message "Program terminated normally" and the report is executed by the debugger.

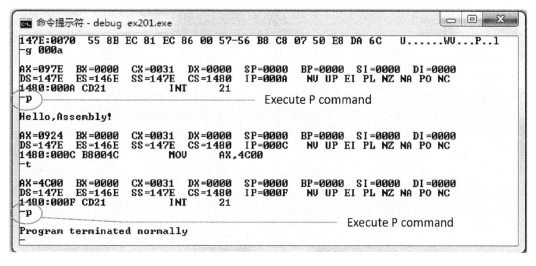

Figure 2-18 DEBUG P command

The execution process command P may be used to execute one or more instructions. The P command is similar to the T command, but is not tracked into the called program because of a subroutine call or a soft interrupt call. The P command is a step-by-step trace, so the LOOP instruction can be executed at a time, or a string operation instruction can be executed once.

Therefore, in this case, we implement the interrupt instruction INT with the P command, don't use T command, under normal circumstances do not enter the DOS function call program and BIOS program.

⑧ Exit DEBUG

To exit DEBUG, simply enter the exit command Q at the DEBUG prompt. After issuing the Q command, DEBUG terminates and the control will go back to the DOS command line. Note that the Q command does not save the file being debugged, and other DEBUG commands must be used to save the contents that are being debugged.

At this point, the debugging process ends in Example 2-1. It is recommended that readers use the DEBUG auxiliary learning assembly instruction and assembly program design in the follow-up study to master some of the commonly used DEBUG debugging commands listed in Table 2-3, which will make the assembly language more intuitive and more profound to understand.

2.4 Assembly language source organization

Corresponding to the segmentation structure of the memory, the source language of the assembly language consists of four basic logical segments: the code segment, the data segment, the stack segment, and the additional segment. Where the code segment is an assembly language program and must have the segment, the other segments may be optional.

2.4.1 Assembly language statements

Although the assembly language statement is simpler than a high-level language statement, it has two completely different statements: one is the instruction and the other is a directive. These two statements are very different. When assembling the source program, the assembler translates the instruction statement into a machine instruction, and the instruction statement has machine instructions that correspond to it. The pseudo-instruction statement does not have a corresponding machine instruction, but only indicates how the assembler assembles the source program, including the definition of the symbol, the definition of the variable, the definition of the logical segment and so on. Thus, instruction statements are associated with a specific processor, independent of the assembler; the directives are independent of the specific processor type, but are related to the assembler, and the pseudo-instructions supported by different versions of the assembler will be different. Instruction statements and pseudo-instruction statements are similar in format, consisting of four parts.

(1) The format of the command statement

```
[Label:] instruction mnemonic [operand [, operand]] [; comment]
```

The instruction mnemonic reflects the function of the instruction, which will be introduced in subsequent chapters.

Operands can be constants (numeric expressions) operands, register operands (register names), or memory operands (address expressions). Whether the instruction has an operand depends entirely on the instruction itself, some instructions have no operands, some instructions have only one operand, some instructions require two operands, and the commas are separated. When there are two operands in the instruction, the operand is called the destination operand, followed by the source operand.

The use of the label depends on the needs of the program. The label is only recognized by the assembler, indicating the address of the instruction, which is independent of the operation of the instruction itself.

The semicolon-based annotations are purely for the purpose of understanding and reading the program, and the assembler ignores all of it and does not affect the instruction. Whether to write a note is determined by the programmer. In order to read and understand the convenience of the program, should be properly used to annotate, by comment to explain the function of the statement or program. Sometimes the whole line can be used as a comment, as long as the line is guided by a semicolon.

(2) The format of the directive statement

```
[ Name ] directive definition character [ parameter, .., parameter ]
   [ ; comment ]
```

The pseudo-instruction delimiter specifies the function of the directive, which will be introduced in subsequent chapters.

General pseudo-instruction statements have parameters for describing the operation of the pseudo-instruction object, the type and number of parameters with the pseudo-instructions vary. Sometimes the parameters are constants (numeric expressions), and sometimes specific symbols.

The names in the directive statements are sometimes necessary and can be saved, which is also related to specific directives. In the assembly language source program, the name and label are easy to distinguish, the name no colon, but after the label it must have a colon.

Usually a statement is used to write a line, the statement between the various components has a separator. After the label of the colon, note guide semicolon and operand, the comma between the parameters are specified by the use of separators. In addition, spaces and tabs are also commonly used delimiters, and the role of multiple spaces or multiple tabs with a space or the role of the same tab, so often in the statement line by adding spaces and tabs method to align the upper and lower lines of the statement to facilitate reading.

The label and name reflect the logical address attributes of the corresponding statement, and also have some attributes of their own. The label and name are programmers' identifiers that conform to the assembly language syntax. Identifiers are typically composed of up to 31 letters, numbers, and special characters (?,@, -, $), and can not start with a number. Labels and names can not be reserved words for assembly language. The reserved words in the assembly language are mainly instruction mnemonics, directive delimiters and register names, and some other special reserved words. It is important to note that by default, the assembler does not distinguish between capitalization of letters. In order to improve the readability of the program, the label and the name should be as meaningful as possible.

2.4.2　Assembly language source program format

A complete assembly language source program can contain several code segments, data segments, additional segments or stack segments, and the order between segments can be arranged arbitrarily. Programs that need to run independently must contain a code segment and indicate the starting point of the program execution, and a program has only one program starting point. All instruction statements must be in a code segment, and the directive statement can be in any section as needed.

In the previous section (Section 2.3), we detailed the development of a simple assembly language source program (Example 2-1) development process, the assembly language program has a preliminary understanding and experience. The source program uses the complete segment definition format. Typically, a complete assembly language source program format mainly includes the following:

① The processor selects the directive.
② Logical segment definition of the directive.
③ Use the setting statement.
④ The program begins.
⑤ The procedure is terminated.
⑥ Assembly end.

[**Example 2-2**]   Assembly language source program complete section defines the typical format.

```
        .8086                    ;(1) the processor selects the directive
        data1 SEGMENT            ;(2) data segment definition (according to need
                                     to set, but also no)
        ...                      ; position A, data variable definition statement
                                     sequence
        data1 ENDS
        data2 SEGMENT            ; data segment definition (according to need to
                                     set, but also no)
        ...                      ; position B, data variable definition statement
                                     sequence
        ...
        data2 ENDS
        Code SEGMENT;
            ASSUME CS: code, DS: data1, ES: data2
                                 ;(3) segment using the setting statement
        start: MOV AX, data1     ;(4) program start point
               MOV DS, AX        ; set segment register
               ...
               ...               ; position C, program body part: instruction
                                     sequence
               ...
               MOV AX, 4C00H     ;(5) end of program, return to DOS
               INT 21H
               ...               ; position D, subroutine part
        subp PROC NEAR           ; defines a subroutine called subp
               ...
               RET
        subp ENDP
               ...
        code ENDS
               END start         ;(6) end of assembly
```

Under normal circumstances, we follow the source program framework to write assembly language source program. The pseudo-statement sequence is arranged at position A (data segment) and can also be placed at position B (additional segment). The most flexible storage of data, of course, can also be in the code segment and stack segment. The main part of the program, the instruction statement part must be placed in position C (code segment). Subroutines are generally arranged at position D, that is, after the end of the code segment program, before the end of the assembly, since the subroutine is executed by the main program call.

1. The processor selects the directive

When you need to use 80x86 series CPU instructions other than 8086, you should indicate the processor used in the first line of the program, called the processor select directive. Such as:

.386 .386P .486 .486P .586 .586P .686 .686P

Among them, .386 that the program selected 80386 basic instruction set, .386P selected 80386 directive set, including privilege command. So on and so forth. The default processor selection directive is .8086, that is, when using only the 8086 instruction system, you can omit the processor selection directive.

2. Logical segment definition directive

The logical segment definition is defined in pairs by using the SEGMENT and ENDS pairs of pseudo-instructions. The segment definition statement is used to organize the program and use the memory in logical segments. Need to match the ASSUME directive to indicate whether the logical segment is a code segment, a stack segment, a data segment, or an additional segment. The general format for a complete logical segment definition is as follows:

```
Segment name SEGMENT [location type] [combination type] [use type]
    ['category']
...        ; statement sequence
...
Segment name ENDS
```

Segment start statement the segment name in SEGMENT is the same as the segment name in the end of the segment ENDS, so that pairing is maintained. The segment name is a programmer-defined identifier.

The segment name can be unique, or it can be the same as other segments in the program. In the same source file, if the segment has been defined with the same segment name, then the current segment is treated as a continuation of the previous segment, that is, the same segment.

For a member of the same name in the source file, the optional value of the subsequent fiducial SEGMENT should be the same as the previous sibling, or the optional value is no longer given the same as the previous sibling.

The segment Start Statement The options "Target Type", "Combination Type", "Usage Type" and "Category" in SEGMENT inform that the assembler and the linker how to create and combine segments. These options are not required, and if given, these options should be described in order, and if no option is given, the assembler uses the default value for that option.

(1) Paragraph positioning (align) attribute

The segment locator attribute specifies the boundary of the logical segment in the main memory, notifying the connecting program how to determine the starting address of the logical segment. Segment location attributes can have the following options:

① BYTE: byte address starts, any address starts (... xxxx xxxxb).
② WORD: word boundary start, even address (... xxxx xxx0b).
③ DWORD: double word boundary start, 4 times the address (... xxxxxx00b).
④ PARA: small segment boundary start, 16 times the address (... xxxx 0000b).
⑤ PAGE: page boundary start, 256 times the address (... 0000 0000b).

The default positioning attribute is PARA.

(2) Section combination (combine) attribute

The same source file allows multiple segments of the same name to appear, which are eventually merged into one segment. If a segment of the same name appears in a different source file, you can specify how the segment is processed with the segment combination attribute when the segment is defined. Segment combination attributes can have the following options:

① PRIVATE: this paragraph does not merge with other modules in the same section, each paragraph has its own segment address. This is the default segment definition of the pseudo-instruction default segment combination.

② PBLBLIC: this paragraph is connected with all other segments of the same type with the same name, synthesize a large physical segment, and specify a common segment address. There is a gap between the original segment which is less than 16B.

③ COMMON: the same name overlap, forming a section, the content is ranked in the final section of the content.

④ STACK: all STACK segments are seamlessly merged in the same way as PUBLIC segments. This is the row segment attribute that the stack segment must have.

(3) Use (use) type attribute

The use of type attribute is to support 32-bit segment and set the properties, only in the use of 80386 above the instruction system assembly program, there are two options (the default type is USE16):

USE16: 16-bit addressing mode, segment length does not exceed 64K.
USE32: 32-bit addressing mode, up to 4GB.

(4) Segment class (class) attribute

A section in addition to a paragraph name, you can also have a category name. The category name is any string that is enclosed in single quotation marks. Classes with the same name are placed in an adjacent storage interval, but are not merged into the same segment. When the join program is organized, all of the same categories are allocated adjacent.

If a segment does not give a category, then the segment's category is empty. Most MASM programs use 'code', 'data', and 'stack' to name the segment of the code snippet, the data segment, and the stack segment respectively to keep all code and data contiguous.

3. Use the setting statement

The assembler determines the logical segmentation of the source program according to the segment start statement SEGMENT and the segment end statement ENDS. In order to effectively generate the object code, the assembler also understands the correspondence between the logical segment and the segment register. The correspondence between the segment register and the logical segment is described by the segment using the set statement ASSUME.

The simple format for using the set statement is as follows:

```
ASSUME segment register name: segment name [, segment register name:
    segment name,…]
```

The segment register names can be CS, DS, SS, and ES. The segment name is the segment name of the segment start statement SEGMENT and the segment end statement ENDS. For example, the following ASSUME statement tells the assembler that the CS register corresponds to the CSEG segment and the DS register corresponds to the DSEG segment from now on.

```
ASSUME CS: CSEG, DS:DSEG
```

The segment name field in the ASSUME directive can also be a special keyword, NOTHING, which indicates that a segment register is no longer associated with any segment.

ASSUME statement can be established in a number of segments and the relationship between the segment, as long as separated by commas. Multiple ASSUME statements can be used in the source program. Normally, the ASSUME statement is used at the beginning of the code segment to determine the correspondence between the segment register and the segment. You can then use the ASSUME statement to change the established relationship as needed. The segment using the set statement ASSUME is a directive statement that can not set the value of the segment register.

4. The program starts

In order to indicate where the program starts execution, a label (such as a start identifier) is required.

During the connection to the source program, the connection program sets the CS and IP values correctly according to the program start point and sets the SS and SP values according to the program size and the stack segment size. The connection program does not set the DS and ES values. If you use a data segment or an additional segment, you must explicitly assign a value to DS or ES.

Most programs require data segments, and the execution of the program is generally:

```
start: MOV AX, data1      ; starting point
       MOV DS, AX         ; set DS
```

5. The program terminates

At the end of the application execution, control should be returned to the operating system. In the assembly language programming, there are a variety of ways to return to DOS, but generally use the DOS function call 4CH sub-function to achieve, it needs the entry parameter AL = return number (usually 0 that the program is not wrong).

Therefore, the application's termination code is generally:

```
MOV AX, 4C00H      ; End of program execution, return to DOS
INT 21H
```

6. End of assembly

The end of the assembly indicates that the assembler completes the process of translating the source code into the target module code, rather than ending the execution of the program. The source program must have an END directive at the end.

```
END [label]
```

The optional "label" parameter is used to specify the execution start point (eg start identifier) corresponding to the program, and the connection program will set the CS:IP value accordingly.

The end of the procedure and the end of the compilation are different concepts, paying attention to distinguishing and understanding their different roles and don't confuse them.

7. Definition of subroutine

It is a program design, which often need to be repeated in different places repeatedly identified as subroutines, subroutines have a certain degree of independence, is to complete a specific function of the program. The definition of a subroutine is not necessary in the assembler source program. If a subroutine is designed, it is important to note that the definition of the subroutine can not be arranged between the program start point and the program termination point. Subroutine is executed by the main program call, so it is generally arranged between the end of the program and before the end of the assembly.

In assembly language, the subroutine is declared with a pair of directives PROC and ENDP, in the following format:

```
Subroutine name PROC [NEAR | FAR]
... subroutine body
Subprogram name ENDP
```

The subroutine name is the symbolic address of the subroutine entry. Subprogram name should be a valid identifier, subroutine name can not be the same source program in the label, variable name, other subroutine name the same. The optional parameter (NEAR/FAR) specifies the calling attribute of the subroutine. When the invocation attribute is not specified, the default attribute NEAR is used.

(1) NEAR: specifies the proximity subroutine. Subroutines and invocations must be in the same code segment (called within the section) and can be omitted.

(2) FAR: designated as remote subroutine, subroutine and caller can be in different code segments (inter-segment calls).

2.5 Operands in assembly language

Assembly instructions The object to be manipulated is called an operand. Often, assembly language recognizes operands with constants, registers, variables, labels, and expressions. The register structure of the 8086/8088 CPU is described in Section 2.2.3. This section describes the rest of the operands.

2.5.1 Constant

A constant represents a fixed value which is a pure value without any attributes. In the compilation phase, its value can be fully determined, and the process of running the program will not change. Constants are available in a variety of forms: constants, strings, numeric expressions, symbolic constants. The numerical expressions and symbolic constants are described separately in sections 2.5.3 and 2.5.4, respectively.

1. Constant

This refers to the values expressed in binary, octal, decimal, and hexadecimal. In the assembly language, a variety of binary data are used to distinguish between the letters.

(1) Binary: a sequence of numbers consisting of 0 and 1, such as 10110110B ending with the letter B or b.

(2) Octal: a sequence of numbers consisting of 0 to 7, such as 125Q, ending with the letter Q or q.

(3) Decimal: a sequence of numbers consisting of 0 to 9, such as 2059D, ending with the letter D or d.

(4) Hexadecimal: a sequence of numbers consisting of $0 \sim 9$ and $A \sim F$ ($a \sim f$) ending with the letter H or h. In order to avoid confusion with identifiers (such as labels, names, reserved words), hexadecimal numbers must begin with a number. So, any hexadecimal number beginning with the letters A to F must be preceded by a 0. Such as 23AFH, 0A25DH.

In the assembly language, the default base is the decimal number, that is, the decimal number in the expression can not add suffix letters, such as: 2059D can also be written as 2059. Assembly language provides a directive to change the base. E. g:

```
.RADIX 16; changes the default radix to hexadecimal
```

2. Strings

A string constant is one or more characters enclosed in single or double quotation marks. The value of the string constant is the ASCII code value of the character enclosed in quotation marks.

For example, the value of 'A' is 41H, the value of 'ab' is 6162H, and the value of 'AB' is 4142H.

3. Numeric expression

The so-called numerical expression in the assembly process and can be calculated by the assembler to calculate the value of the expression. The parts that make up the numerical expression must be fully determined at assembly time.

4. Symbolic constants

The symbolic constant uses a identifier to express a value. The directives for defining symbol constants are "EQU" and "=".

2.5.2 Variables and labels

Variables and labels represent storage units. The variable is stored in the storage unit, and the instruction code is stored in the storage unit indicated by the label. The definition of the label is very simple, it is an optional part of the instruction statement, see Section 2.4.1 Assembly statement.

The address of the memory cell represented by the variable is unchanged, but the data stored therein can be changed. Variables need to be defined before they can be used. After the definition of the variable can be used variable name and other methods of its representative of the memory unit data, that is, the value of the variable.

1. Definition of variables

The variable definition directive applies the storage space in fixed length for the variable, and the corresponding storage unit can be initialized.

The variable-defined assembly language format is:

```
[Variable name] variable definition directive parameter [,…, parameter]
   [; comment]
```

The variable name is a programmer's custom identifier, and the parameter consists of a constant or "?". Where "?" indicates that the assigned memory cell is not assigned an initial value. The variable name is optional, it directly represents the first storage unit in the allocated storage space defined by the statement. E. g:

```
BUFF DB 100, 12, 56      ; BUFF represents the first byte storage unit,
                           which stores the initial value of 100.
```

Multiple storage units, if the initial values are the same, you can use the copy operator (repeat delimiter) DUP to define:

Number of repetitions DUP (repeat parameters)

E. g:

```
BUFF DB 3 DUP (12), 56
```

The variable defines the pseudo-instruction find allocate 4 bytes of memory cells, the first three units have the same initial value of 12, the latter unit initial value of 56.

Variable definition directives are: DB(Define Byte), DW(Define Word), DD (Define Double Word), DQ (Define QuarWord), DT (Define Ten Byte) and so on. Their functions are shown in Table 2-5. In addition to DB, DW, DD there are other simple variables defined. The assembly language also supports complex data variables, such as structure (Structure), record (Record) and so on.

Table 2-5 Variable definition directives

Mnemonic	Variable type	Function of defining variables
DB	byte	Apply the main memory space for each data in bytes (8b)
DW	word	Apply the main memory space for each data in units of words (16b)
DD	Dword	Apply the main memory space for each data in double word (32b)
DQ	8 bytes	The main memory space is requested for each data in 8-byte (64b) units
DT	10 bytes	Apply the main memory space for each data in 10 bytes (80b)

2. Application of variables

Defined variables can be referenced by methods such as variable names. Variable names can be used in instruction statements, or can be used in pseudo-statement statements, combined with Example 2-3 attention to understand their differences.

① Use in the instruction statement: you can refer to the first data pointed to by the variable name, through the variable name addition and subtraction displacement access to the first data as the base address before and after the data. In the instruction statement directly reference variable name, which means access to the variable name points to the storage unit.

② Use in pseudo instruction statement: mainly by another variable reference, that is, in another variable definition statement as part of the parameters appear. Note that the variable name appears as a separate parameter in the variable definition directive, it can only appear in the DW or DD directive statement, can not appear in the DB, DQ and other variable definition statements.

[**Example 2-3**] Definition and application of variables.

```
;Source: ex203.asm
Data segment
DSEG SEGMENT
BVAR DB 1, -2, 'AB', 3 DUP ('a'),?        ; Byte variable, 8 items
WVAR DW 1, -2, 'AB', 3 DUP ('a'),?        ;Word variable, 7 items
DVAR DD 1, -2, 'AB', 3 DUP ('a'), double word variable, 7 items
VAR1 DW WVAR, DVAR, DVAR-WVAR, VAR1-DVAR  ;word variable, 4 items
VAR2 DD WVAR, DVAR, DVAR-WVAR, VAR1-DVAR  ;double word variable, 4 items
VAR3 DB DVAR-WVAR, VAR1-DVAR              ;byte variable, 2 items
DSEG ENDS
;Code snippet
MOV CL, BVAR                    ;BVAR The first data is sent CL,
                                 (CL) =01H
MOV CH, BVAR+2                  ;BVAR The third data is given CH,
                                 (CH) =41H
MOV BX, WVAR                    ;WVAR The first data is sent to
                                 BX,(BX) =0001H
MOV SI, WVAR+2                  ;WVAR The second data is sent SI,
                                 (SI) =0FFFEH (-2)
MOV DX, WORD PTR DVAR           ;DVAR The first data is low to
                                 send DX, (DX) =0001H
MOV AX, WORD PTR DVAR+2         ;DVAR first data high word to
                                 send AX, (AX) =0000H
MOV DI, VAR1+4                  ;VAR1 The third data is sent to
                                 WVAR 7th data unit.
MOV WVAR+12, DI
```

The above definition of the variable definition statement corresponding to the storage area allocation (Figure 2-19), the figure in the digital value in hexadecimal representation. Observe the use of different variables to define the directive, the same data stored in memory. For example: 'AB'.

Offset address	Content	Variable name
0000H	01H	BVAR
0001H	0FEH	
0002H	41H	
0003H	42H	
0004H	61H	
0005H	61H	
0006H	61H	
0007H	00H	
0008H	01H	WVAR
0009H	00H	
000AH	0FEH	
000BH	0FFH	
000CH	42H	
000DH	41H	
000EH	61H	
000FH	00H	
0010H	61H	
0011H	00H	
0012H	61H	
0013H	00H	
0014H	00H	
0015H	00H	
0016H	01H	DVAR
0017H	00H	
0018H	00H	
0019H	00H	
001AH	0FEH	
001BH	0FFH	
001CH	0FFH	
001DH	0FFH	
001EH	42H	
001FH	41H	
0020H	00H	
0021H	00H	
	...	

Figure 2-19 Example 2-3 variable definition storage assignment

The value of each register after the program is exclusive as shown in Figure 2-20.

CHAPTER 2　ASSEMBLY LANGUAGE BASICS

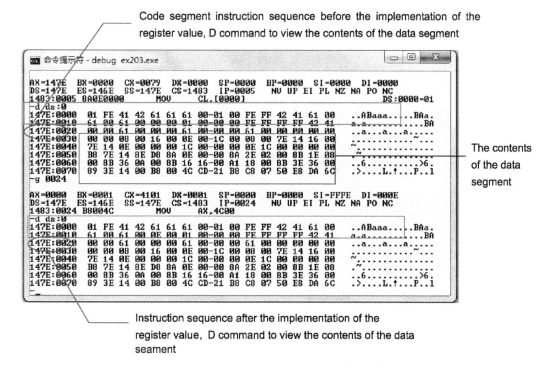

Figure 2-20　Example 2-3 debugging results

Note: variable name addition and subtraction displacement of the representation, there can be a variety of forms, the following forms are equivalent:

① MOV CH, BVAR + 2
② MOV CH, BVAR [2]
③ MOV CH, [BVAR + 2]

3. The positioning of variables

The assembler determines the starting address of each logical segment according to the boundary location attribute specified by the segment definition instruction according to the sequential order of the instruction in the order in which the instruction is defined. For example, Example 2-3 variable definition in turn from the data segment offset address of 0 units began to one by one distribution. The locating directive is provided in the assembly language to change this default. The instruction format is as follows:

```
ORG parameter        ; the parameter value will be the offset address of
                       the next instruction statement or variable
```

E. g:

```
DATA SEGMENT
    ORG 10H
    VAR1 DB 1, 'A'
    ORG $ +2
    VAR2 DW 1234H, $ - VAR1
DATA ENDS
```

059

Where the operator '$' represents the current offset address value, which is the offset address of the next allocated memory cell. The distribution of VAR1 and VAR2 in memory is shown in Figure 2-21.

Offset address	Content	Variable name
0010H	01H	VAR1
0011H	41H	
0012H	−	
0013H	−	
0014H	34H	VAR2
0015H	12H	
0016H	06H	
0017H	00H	
	⋮	

Figure 2-21 VAR1 and VAR2 variables define storage allocation

4. Attributes of variables and labels

Both variables and labels represent storage units. The variable is stored in the storage unit, and the instruction code is stored in the storage unit indicated by the label. So, variables and labels have the following three attributes:

(1) Segment attribute (SEG): the variable or label corresponds to the segment address of the segment where the storage unit is located.

(2) Offset address attribute (OFFSET): variable or label corresponding to the first storage unit within the offset address.

(3) Type attribute (TYPE): variable type attribute refers to the variable which occupies the storage unit bytes. The attribute value is determined by the variable definition directive. The main variable and label type attributes and return values are shown in Table 2-6.

Table 2-6 Types of memory operands and return values

type	DB	DW	DD	DF	DQ	DT	NEAR	FAR	constant
value	1	2	4	6	8	10	−1	−2	0

In assembly language programming, the three attributes of variables and labels are important. Assembly language provides a special value of the operator and the type of operation on the variables and labels of the three attributes of the corresponding operation and processing.

5. Value of the operator

The value evaluators are also called numeric loopback operators because these operators send some of the features or memory addresses as numeric values. The value of the operators are SEG, OFFSET, TYPE, SIZE and LENGTH, etc., their use of the format and role are shown in Table 2-7.

CHAPTER 2   ASSEMBLY LANGUAGE BASICS

Table 2-7   commonly used value operator

Value of the operator	Function
SEG Variable name/label	Returns the segment address of the variable or label
OFFSET Variable name/label	Returns the offset address of the variable or label
TYPE Variable name/label	Returns the type of the variable or label, the type is represented by a value
LENGTH Variable name	Returns the number of elements in the variable defined by the DUP, that is, the number of repetitions before the DUP, and returns 1 in other cases
SIZE Variable name	Return value of LENGTH × TYPE

6. Attribute operator

In order to improve the flexibility of the access variables, labels and general memory operands, the assembly language also provides the attribute operators PTR and THIS, etc., to achieve the purpose of the specified attribute access. The format and function are shown in Table 2-8.

Table 2-8   Common Attribute Operators

Attribute operator	Function
Type PTR variable name/label	Temporarily assign or temporarily change the use of variables and labels
THIS type	Used to create an operand with the current address but for the specified type
SHORT label	The label is treated as a short transfer
Segment register:	Used to specify a segment attribute for a memory operand, that is, a segment that goes beyond it

Note: "Type" can be BYTE, WORD, DWORD, FWORD, QWORD, TBYTE, NEAR and FAR, etc., can also be defined by the structure, records and other types.

[**Example 2-4**]   Attribute and its application, data segment variable definition of the same case 2-3.

```
; Source: ex204.asm
; Code snippet
    MOV   BP, WORD PTR BVAR
    MOV   DX, WORD PTR BVAR + 2
    MOV   BL, BYTE PTR WVAR
    MOV   SI, SEG WVAR + 2
    MOV   ES, SI
    MOV   SI, OFFSET WVAR + 2
    MOV   BH, BYTE PTR DVAR + 8
    MOV   DI, WORD PTR DVAR + 8
    MOV   AL, TYPE WVAR
    MOV   AH, LENGTH WVAR
```

The value of each register after the program is executed is shown in Figure 2-22.

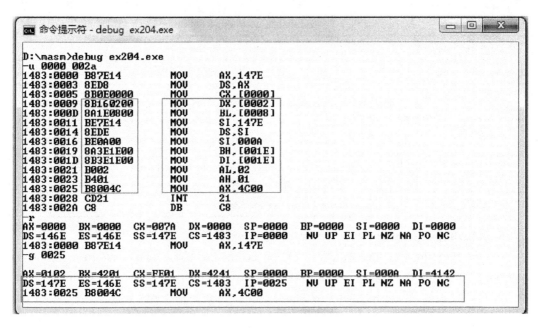

Figure 2-22　Example 2-4 debugging results

2.5.3　Expression

Expressions are one of the common forms of operands, and are concatenated by operators, such as constants, variables, or labels. In assembly language, the expression is divided into numerical expressions and address expressions. The value of the address expression is the address of a memory in which data (called a variable) or an instruction (called a label) is stored. The value of a numeric expression is determined by the assembler in the compilation phase, rather than it been calculated during the run of the program. Therefore, the parts of the numerical expression must be determined at the time of assembly, and finally a definite value is obtained, so the numerical expression is also a constant.

Assembly language supports a variety of operators, as shown in Table 2-9.

Table 2-9　MASM supported Operators

Operator type	Operation symbols and descriptions
Arithmetic operator	＋(Plus), －(minus), * (multiply), / (except), MOD (take)
Logical operators	AND, OR, XOR, NOT, NOT
Shift operator	SHL (logical left shift), SHR (logical right shift)
Relational operator	E (equal), NE (unequal), GT (greater than), LT (less than), GE (greater than or equal to), LE (less than or equal)
Value of the operator	OFFSET, SEG, TYPE, LENGTH, SIZE
Attribute operator	:, PTR, THIS, HIGH, LOW, SHORT
Other operators	( ), [ ], ·, < >, MASK, WIDTH

An expression is a combination of a constant, a variable, a label, and an operator. If an expression has more than one operator, the following rules are used:

(1) The first high priority operation, the lower the priority operation.

(2) The same priority, according to the expression from left to right in the order of operation.

(3) Brackets can increase the priority of the operation, and the operations in parentheses are always done before the adjacent operations.

The order of the precedence of the various operators is shown in Table 2-10. The operators of the same row in the table have equal priority, priority 1 is the highest level, and priority 10 is the lowest level. The value of the expression is calculated by the assembler, and the program is written to the correct order, so that the program is wrong. Try to use parentheses to specify the order of operations.

Table 2-10 Operation priority order

priority	Operator
1	( ),[ ],·,< >,MASK,WIDTH,LENGTH,SIZE
2	PTR,OFFSET,SEG,TYPE,THIS,CS:,DS:,ES:,SS:
3	HIGH,LOW
4	*,/,MOD,SHL,SHR
5	+,-
6	EQ,NE,GT,LT,GE,LE
7	NOT
8	AND
9	OR,XOR
10	SHORT

[**Example 2-5**] Numerical expression and its application.

```
; Assembly language instructions after the assembly, the expression of the
  formation of instructions
  MOV   AX, 3 * 4 + 5              ; MOV  AX,17
  MOV   DH, 01100100B SHR 2        ; MOV  DH,19H
  MOV   BL, 8CH AND 73H            ; MOV  BL,0
  MOV   AH, 8CH OR 73H             ; MOV  AH,0FFH
  MOV   AX, 10H GT 16              ; MOV  AX,0
  MOV   BL, 6 EQ 0110B             ; MOV  BL,0FFH
  MOV   BX, 32 + ( ( 13 /6 ) MOD 3 )  ; MOV  BX,0022H
```

2.5.4 Symbol definition

In assembly language, sometimes the same numerical or expression occurs many times. For convenience, it can be given a symbol by a symbolic definition directive, which can then be used instead of the value or expression. Define symbols used to represent constants, strings, or numeric expressions through symbolic definition statements, which are also called symbolic constants. Because constants, strings, numeric expressions are constants.

Commonly used symbol definitions are: EQU, =(equal sign) and LABLE.

1. Equivalent statement EQU

The general format of the equivalent statement is as follows:

```
Symbolic name EQU expression
```

The symbol name is a custom identifier for the programmer; the expression can be a constant, a symbol, a string, a numeric expression, or an address expression.

E. g:

```
CR EQU 0DH                      ; constant
LF EQU 0AH
NUM EQU 4 * 128                 ; numeric expression
ADR EQU ES:[BP + DI +5]         ; Address expression
MOVE EQU MOV                    ; instruction mnemonic
HELLO EQU "Hello, Assembly!"    ; String
```

When assembling, the symbol names defined for EQU are "replaced" with the corresponding expression.

E. g:

```
MSG DB HELLO ; Equivalent to MSG DB "Hello, Assembly!"
MOVE CX, NUM + 1Equivalent to MOV CX, 4 * 128 + 1
```

Using the EQU directive, you can use a name to represent a value, or use a short name instead of a longer name. If you need to reference an expression multiple times in the source program, you can use the EQU pseudo-operation to assign a name to replace the expression in the program, making the program more concise and easy to read. In the future, if you change the value of the expression, you have to modify one, the program is easy to maintain.

Note:

① The EQU directive does not assign a symbol to the memory cell, noting that it differs from the variable definition directive.

② The symbol specified by the EQU directive can not be the same as another identifier or keyword, nor can it be redefined. Otherwise, the assembler considers that the symbol redefines the error.

③ If you want to use the symbol in the program, you must follow the "first defined after the use of" rules.

2. Equal sign(=)

The assembly language also provides an equal sign to define a symbolic constant, that is, a symbol or a numeric expression. The general format of the equal sign is as follows:

```
Symbolic name = numeric expression
```

E. g:

```
X = 10
Y = 20 + 300 /4
```

The equal sign function is similar to EQU, with the main difference being that =(equal sign) can be repeated for the same symbol.

E. g:

```
X = 2 * X + 10          ; but X EQU 2 * X + 10 is wrong
```

3. Define symbolic statements (LABEL)

Defining a symbolic name statement (LABLE directive) is the type that defines the label or variable, which shares the memory unit with the next instruction. The general format is as follows:

```
Symbol name LABEL type
```

Types can be BYTE, WORD, DWORD, NEAR and FAR. The function of this statement is to define the symbol specified by the "symbolic name" whose segment attribute and offset attribute are the same as the segment attribute and offset attribute of the next immediately following storage unit, the type of which is the parameter "type", the type specified. Using the LABLE directive allows the same data area to have both types of attributes. Such as: BYTE (byte) and WORD (word), so that in the future of the program according to the different needs of the byte or word units to access the data.

E. g:

```
VARW LABLE WORD          ; Variable VARW type is WORD
VARB DB 6 DUP (?)        ; Variable VARB type is BYTE
...
MOV VARW, AX             ; AX in the first and second bytes
...
MOV VARB [4], AL         ; AL in the fifth byte
```

VARW type is WORD, and VARB shared memory, that is, paragraph attributes and offset attributes and VARB the same.

The LABLE directive can also define a label with a property NEAR as FAR.

E. g:

```
QUIT LABEL FAR
EXIT: MOV AX, 4C00H
```

This instruction "MOV AX, 4C00H" has two tags: QUIT and EXIT. But they are of different types.

2.6 8086/8088 addressing mode

The method of finding the operand in the instruction is called the addressing mode. Operators in the computer storage mainly have the following four cases:

(1) The operand is in the instruction area (code segment), that is, the operand is included in the instruction. As long as the instruction is fetched, the operand immediately following the instruction opcode can be obtained. This operand is called immediate.

(2) The operand is located in an internal register of the CPU. The operand field in the assembler instruction is the register name. The data stored in the corresponding register is the operand to be searched. This operand is called the register operand.

(3) The operand is located in a cell in the memory data area or stack area. The operand field in the assembly instruction gives the address information of the memory cell in some way. The operand can be found by knowing the address of the memory cell. The operand is called a memory operand.

(4) The operand is located in the I/O port. The operand field in the assemble instruction gives the address of the I/O port directly or indirectly, as long as the I/O port address is known to find the I/O port operand.

In the 8086/8088 system, the addressing modes of the operands are addressed immediately, register addressing, memory addressing, and I/O port addressing, depending on where the operands are located in different places in the computer. Wherein the address of the memory cell is composed of two parts: the segment address and the intra-segment offset address (EA). Since the bus interface unit BIU can automatically refers to the corresponding segment register to obtain the segment address as needed, the memory addressing mode is mainly to determine the storage unit effective address EA. And then use the "physical address PA = segment address ×16D + EA" to get the physical address of the operand. The valid address EA is a 16-bit unsigned number. EA is constructed in a variety of ways, thus forming a variety of memory addressing modes: direct addressing, register indirect addressing, register relative addressing, base index addressing, and relative base index addressing.

When not specified, the general default memory operand is accessed in the DS segment. If the BP register is used in the addressing mode, the SS segment is accessed by default. The default case allows the use of segments in the instruction to override the prefix change.

The following are the seven basic addressing modes supported by the 8086/8088 system: immediate addressing, register addressing, direct addressing, register indirect addressing, register relative addressing, base index addressing, relative base address, and indexed addressing. In the learning process you should pay attention to the characteristics of the addressing mode and the location of the operand. The I/O port addressing is described in Chapter 6 of this book.

2.6.1 Immediate addressing

The operands that appear directly in the instruction are called immediate addressing. The immediate addressing mode is characterized by the immediate number as part of the instruction, immediately following the instruction code of the instruction, stored in the code segment of the memory; the immediate number can be an 8-bit or 16-bit integer that appears as a constant, The immediate data can be observed in the code.

E. g:

```
MOV BH,12H
MOV DX,2012H
MOV AX,'A'
MOV BL,'B'
```

The above instruction source operand is the immediate addressing mode, 12H, 2012H, 'A','B' are constant. The addressing procedure for the 4-instruction source operand is shown in Figure 2-23. The debug result is shown in Figure 2-24. (In this case into the debug command, followed by the use of the U, R, G command, why? For other addressing methods, the reader can also use DEBUG to help learn to understand.)

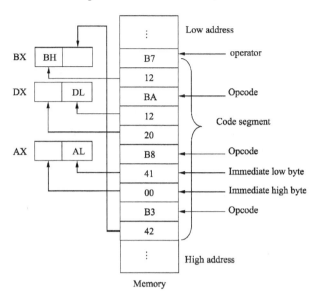

Figure 2-23   Direct addressing

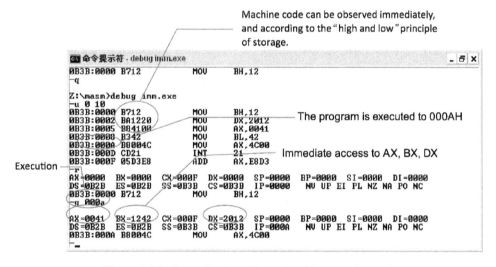

Figure 2-24   Immediately address the debug result graph

Among them, the character A represents 0041H, and the character B represents 42H, because they must meet the principle of data type match. In the instruction "MOV AX, 'A'", the destination operand AX is a 16-bit register, so the character A is automatically expanded to 16 bits (0041H) by an 8-bit ASCII code value (41H). In the instruction "MOV BL, 'B'", the destination operand BL is an 8-bit register, which matches the 8-bit ASCII code of the character B without extension.

The immediate number can only be used as a source operand in the instruction. The immediate addressing mode is mainly used to assign an initial value to a register or memory location, except for the segment register and flag register. In order to transfer data to the segment register, the immediate data should be assigned to a general register and then transferred from the general register to the segment register.

E. g:

```
MOV AX,1234H        ; AX←1234H
MOV SS,AX           ; SS←(AX)
```

2.6.2 Register addressing

Operands are stored in a register inside the CPU, which is called register addressing. The register addressing mode is characterized by the register name appearing in the assembler instruction, which can be 8-bit and 16-bit.

E. g:

```
MOV BX, AX
MOV DH, CL
```

The above instruction source operands and destination operands are immediately addressed.

Since the register operand is located inside the CPU, the addressing process does not involve bus operation, so the register addressing mode is faster. The source operand and destination operand in a instruction can be addressed by using a register, but both must be equal. Note that register addressing can not use both IP and PSW registers.

2.6.3 Direct addressing

The storage unit valid address EA storing the operands is contained in the instruction, that is, EA is provided directly by the instruction, which is called direct addressing. The direct addressing mode is characterized by the fact that the EA given in the assembly instruction is a constant enclosed in square brackets.

E. g:

```
MOV AX, [0300H]
MOV CL, [0300H]
```

The above instruction source operand is direct addressing mode. Figure 2-25 gives the first part of the instruction source operand addressing process, two instructions debugging verification process shown in Figure 2-26.

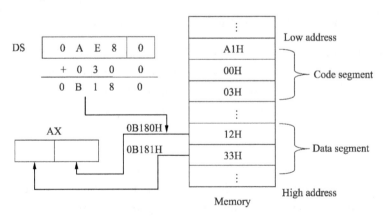

Figure 2-25　Direct addressing

CHAPTER 2 ASSEMBLY LANGUAGE BASICS

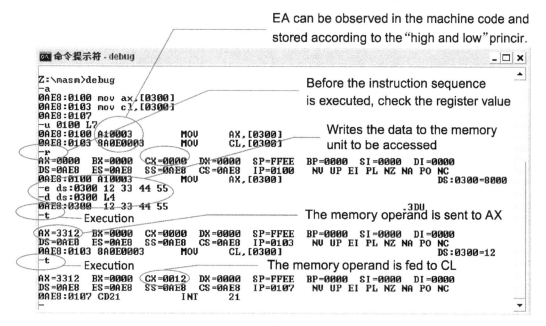

Figure 2-26 Direct addressing debug verification results

Note that the difference between the direct addressing mode and the immediate addressing mode can be seen from the form of the assembler instruction. In the direct addressing instruction, the 16-bit number representing the valid address must be enclosed in square brackets. The function of the instruction completion is not to transfer the constant 0300H To the accumulator AX. After the instruction is executed, the contents of AX are 3312H.

If not specified, the operand of the direct addressing mode is the data segment of the memory, ie the implied segment register is DS, but the 8086/8088 allows the segment to go beyond, allowing CS, SS or ES as the segment register.

E. g:

    MOV BX,ES:[0300H]

This instruction indicates that the ES is a segment register and the operands are addressed in the additional data segment.

2.6.4 Register indirect addressing

The memory address of the storage unit is stored in a register in the BX, BP, SI, and DI of the CPU. This addressing mode is called register indirect addressing. The register indirect addressing mode is characterized by the presence of enclosures in the assembly instructions.

E. g:

    MOV AX,[BX]
    MOV CX,[BP]

The above instruction source operand is the register indirect addressing mode, the second instruction operand addressing process is shown in Figure 2-27.

069

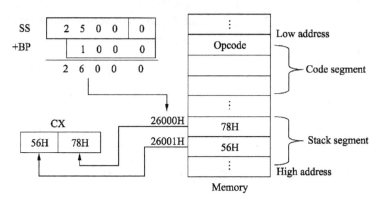

Figure 2-27　BP register indirect addressing

Note that register indirect addressing can only use BX, BP, SI, DI these four registers, other registers are not allowed to use. When using the BX, SI, DI registers, the implied segment register is DS; when the BP register is used, the implied segment register is SS. Allow the use of segment override instructions to change the default.

2.6.5　Register relative addressing

The operand address of the storage unit is the sum of the 8-bit or 16-bit displacement given in the instruction in the contents of one of the BX, BP, SI, and DI registers in the CPU. This addressing mode is called register relative addressing. The relative addressing mode of the register is characterized by the presence of a register enclosed in square brackets in the assembly instruction plus an 8-bit or 16-bit offset, which is the effective address of the operand.

E. g:

```
MOV AX,[SI+1200H]
MOV BX,[BP+2100H]
```

The above instruction source operand is register relative addressing mode, in which the first instruction operand addressing process is shown in Figure 2-28.

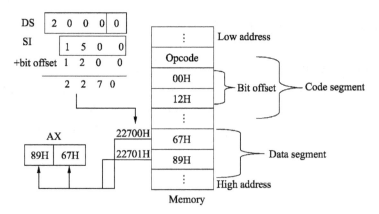

Figure 2-28　SI register relative addressing

Note that the relative addressing of registers can only use the four registers BX, BP, SI, DI, and other registers are not allowed. When using the BX, SI, DI registers, the implied segment register is DS; when the BP register is used, the implied segment register is SS. Allow the use of segment override instructions to change the default. This is the same as the register indirect addressing mode.

When writing assembly language instructions, the register relative addressing can be in several different forms. For example, the following three kinds of writing can achieve the same function.

```
MOV AL,[BP+disp]
MOV AL,[BP]+disp
MOV AL,disp[BP]
```

2.6.6 Base indexed addressing

The operand address of the storage unit that stores the operand is the sum of the contents of the specified address register BX or BP plus the contents of the specified index register SI or DI, which is called base index addressing. The base address indexing mode is characterized by the presence of a base address register and an index register enclosed in square brackets in the assembler instruction. The sum of the contents of the two registers is the valid address of the operand.

E. g:

```
MOV AX,[BX+DI]        ; can also be expressed as MOV AX,[BX][DI]
```

The above instruction source operand is the base address indexing mode. The instruction addressing process is shown in Figure 2-29.

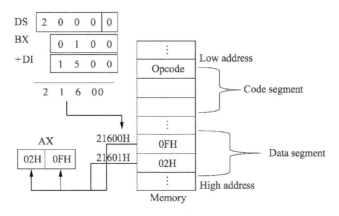

Figure 2-29 Base indexed addressing

Note that the base index addressing must be a combination of base address register BX or BP and an index register SI or DI, which is not allowed both as a base register or as an index register at the same time. As for the implied segment register, it is usually determined by the base register used. When BX is used, the implied segment register is DS. When BP is used, the implied segment register is SS. Allow the use of segment override instructions to change the default.

2.6.7 Relative base indexed addressing

The effective address of the memory cell EA is the sum of the contents of the specified base register (BX or BP) and the contents of the specified index register (SI or DI), plus the 8-bit or 16-bit shift in the instruction (disp), this addressing mode is relative to the base index addressing. The relative address indexing scheme is characterized by the presence of a base register enclosed in square brackets and an index register in the assembly instruction, plus an 8-bit or 16-bit offset, two registers. The sum of the contents plus the sum is the effective address of the operand.

E. g:

MOV AX, 1100H [BX] [DI]

The above instruction source operand is the relative base index addressing mode, the instruction operand addressing process is shown in Figure 2-30.

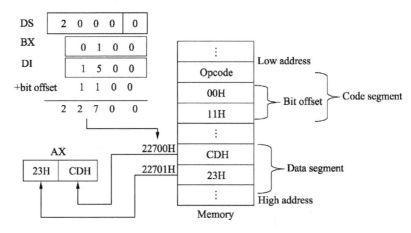

Figure 2-30 Relative base index addressing

Note that the relative base indexed addressing must be a combination of base address register BX or BP and an index register SI or DI, which is not allowed for both the base register or both the index register. As for the implied segment register, it is usually determined by the base register used. When BX is used, the implied segment register is DS. When BP is used, the implied segment register is SS. Allow the use of segment override instructions to change the default. This is the same as the base indexing addressing mode.

When writing assembly language instructions, the relative base index addressing mode can be in several different forms. For example, the following six kinds of wording can achieve the same function:

```
MOV AL,disp[BP][SI]
MOV AL,[BP+disp][SI]
MOV AL,[BP+SI+disp]
MOV AL,[BP]disp[SI]
MOV AL,[BP+SI]disp
MOV AL,disp[SI][BP]
```

[**Example 2-6**] Word array WVAR is defined as follows, please write the sequence of instructions, in a variety of ways to achieve the fourth word element (ie-3) to the AX.

CHAPTER 2 ASSEMBLY LANGUAGE BASICS

```
WVAR DW 1, -2,2, -3,3, -4        ; word variable
; Method 1, direct addressing:
MOV AX, WVAR+3*2                 ; WVAR represents the first element of
                                   the array address, the fourth
                                   element displacement of 3*2
; Method 2, register indirect addressing:
MOV BX, OFFSET DWAR+3*2          ; Array WVAR 4th element offset address
                                   Load BX
MOV AX, [BX]
; Method 3, register relative addressing:
MOV SI, 3*2                      ; WVAR array The fourth element from the
                                   first element of the displacement of
                                   3*2, into the SI
MOV AX, DWAR[SI]
; Method 4, register relative addressing:
LEA BX, DWAR                     ; WVAR array The first element address
                                   is placed in BX
MOV AX, 3*2[BX]
;Method 5, base index addressing:
LEA BX, DWAR                     ; WVAR array The first element address
                                   is placed in BX
MOV SI, 3*2                      ; the fourth element from the first
                                   element of the displacement of
                                   3*2,into the index register SI
MOV AX, [BX+SI]
; Method 6, relative to the base address index addressing: (the offset is
  split into two components to construct, this method is only to show the
  use of relative base index addressing, in this case the method is not very
  appropriate)
LEA BX, DWAR                     ; WVAR array The first element address
                                   is placed in BX
MOV SI, 4                        ; the third element from the first
                                   element of the displacement of
                                   2*2, into the index register SI
MOV AX, [BX+SI+2]                ; the displacement of the fourth
                                   element from the second element
                                   is 2, as the relative amount.
```

Note: the Example 2-6 experiences the different memory addressing mode in the use and set the difference. For the operands in memory, in the program design, one of the addressing modes can be selected as needed.

2.7 8086/8088 instruction system

The microprocessor completes the specified operation by executing the instruction sequence, and the set of all instructions that the processor can execute is the instruction system of the processor. Intel 8086/8088 CPU 16-bit basic instruction set and 32-bit 80 × 86, including the Pentium series is fully compatible, therefore, 8086/8088 instruction system is the basis of the entire Intel 80 ×86 series instruction system. Most instructions can handle word data and handle byte data. The 8086/8088 instruction system can be divided into six functional groups by function: data transmission, arithmetic, logic, string manipulation, program control, control (processor control).

For each instruction, the study should pay attention to the following points:
(1) The assembler format of the instruction;
(2) The function of the instruction;
(3) Instruction to support the operand addressing mode;
(4) The effect of the directive on the mark;
(5) Instructions implied operands and other special requirements.

Special Note: in order to facilitate the introduction of the instruction system, Table 2-11 lists some of the symbols used in this book.

Table 2-11 Operand Symbol Conventions

	Symbols and meanings
Operands	opr:operand. src:source operand. dst: Destination operand
Register addressing	seg:segment register. reg:general register, example reg8 or reg16. reg8:8-bit general purpose register. reg16:16-bit general purpose register
Memory addressing	mem:memory unit, example m8 or m16. m8:8-bit memory cell. m16: 16-bit memory cell
Immediate number	imm: immediate, representative imm8 or imm16. imm8: 8-bit immediate. imm16: 16-bit immediate

2.7.1 Data transfer class instructions

The data transfer instruction can be divided into general transmission instruction, accumulator dedicated transfer instruction, address transfer instruction, flag transfer instruction, lookup table instruction and symbol extension instruction. In addition to the SAHF and POPF instructions, this set of instructions has no effect on each flag.

1. General data transfer instructions

The general transfer instruction includes the basic transfer instruction MOV, the stack instruction PUSH and the POP, the exchange instruction XCHG.

(1) Basic transfer instruction MOV

The MOV instruction is the most frequently used instruction.

◇ Instruction format and function:

```
MOV dst,src        ;(dst)←(src)
```

◇ Supported addressing modes:

```
src: reg,seg,mem,imm
dst: reg,seg,mem
```

The MOV instruction passes the contents of the source operand src to the destination operand dst. When the MOV instruction is executed, the source operand src and the destination operand dst are the same, and the data is actually copied. The MOV instruction can be used for byte data transfer and word data transfer, but src and dst must be equal. And both can not operate for the memory at the same time. The correct data transfer direction for the MOV instruction is shown in Figure 2-31.

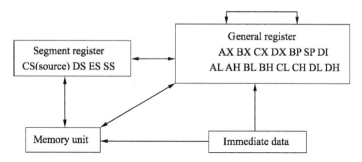

Figure 2-31 MOV instruction data transmission direction

◇ MOV instruction use Note:

① src, dst must have the same type, that is, with the byte type or the same type.

② src, dst can not simultaneously for the memory operand mem.

③ src, dst can not simultaneously for the segment register seg.

④ Immediate imm can not be passed directly to the segment register seg.

⑤ CS can only be used as a source operand and can not be used as a destination operand.

⑥ Immediate imm no address, it can only be used as the source operand, can not be used as the destination operand.

[**Example 2-7**]   Data transmission.

```
MOV CL,05H          ; reg8←imm8
MOV [BX],2008H      ; mem16←imm16
MOV SI,BP           ; reg16←reg16
MOV DS,AX           ; seg←reg16
MOV BX,ES           ; reg16←seg
MOV [BP],ES         ; mem16←seg
MOV AX,[2008H]      ; reg←mem16
MOV [1234H],BX      ; mem16←reg
```

[**Example 2-8**]   The contents of one word of the storage unit 3000H are sent to the 4000H unit.

```
MOV AX, [3000H]     ; can also be transmitted via other general-purpose
                      registers
MOV [4000H], AX     ; but can not be transferred directly between two
                      storage units
```

[**Example 2-9**]   To determine the command right and wrong, and pointed out the cause of the error.

```
MOV CS, AX          ; wrong, CS can not be used as destination operand
MOV AX, CS          ; right
MOV SS, SP          ; right
```

```
    MOV DS, CS         ; wrong, two operands can not be a segment register at the
                         same time
    MOV AX, BL         ; wrong, two operand types are inconsistent
    MOV 30H, AL        ; error, immediate data can not be used as the destination
                         operand
    MOV AL, 300        ; Wrong, source operand is outside the scope of the
                         destination operand
    MOV [DX], BL       ; wrong, DX can not be used for register indirect
                         addressing
    MOV [BX], 30H      ; wrong, two operand types are not clear
    MOV [BX], [100H]   ; Error, two operands can not be both memory operands at
                         the same time
```

(2) Stack operation instructions PUSH and POP

Stack operation instructions are two: the stack instruction PUSH and the stack instruction POP, used to complete the stack and stack operation.

In the 8086/8088 system, the stack is a RAM area. The main memory area where the stack is located is the stack segment. Used to store some temporary data. For example, the entry parameters, return addresses, and so on when calling subroutines. The use of the stack segment and the data segment, the code segment is different, the stack section has the following characteristics:

① Allocates and uses from larger addresses (data segments, code segments are allocated and used from smaller addresses).

② Segment address stored in the SS, SP at any time point to the top of the stack, tracking the top of the stack changes.

③ Stack operation always follow the "first in the first out" principle.

④ Data into and out of the stack are in word units. All data is stored and removed at the top of the stack. Automatically modify SP after entering and leaving the stack.

For example, the following stack section is defined in a program:

```
    SSEG SEGMENT STACK      ; stack section begins
        DW 100 DUP (?)      ; Size is 100 words
    SSEG ENDS               ; stack section ends
```

When the program is loaded, the operating system places the SSEG segment address into the SS, and the number of bytes in the stack segment is placed in the SP, that is, 200 (0C8H), and is allocated from the larger address. When a data is pushed in, the stack top position pointed to by the SP moves toward the low address.

1) Push command PUSH

◇ Instruction format and function:

```
    PUSH src           ; (SP) ← (SP) -2,
                       ; ((SP) +1, (SP))←(src)
```

◇ Supported addressing modes:
    src: reg16, seg, mem16

2) The stack instruction POP

◇ Instruction format and function:

```
POP    dst              ;(dst)←((SP)+1,(SP)),
                        ;(SP)←(SP)+2
```

◇ Supported addressing modes:

```
dst:reg16,seg,mem16
```

◇ PUSH and POP instructions use note:

① The operands of the PUSH and POP instructions must be word types.

② The operands of the PUSH and POP instructions can not be literal imm.

③ SP at any time point to the top of the stack, into the stack stack operation automatically modify the SP.

④ The operand of the POP instruction can not be CS.

⑤ PUSH and POP instructions are reversed operation, in the programming of the two instructions should be used in pairs to achieve "stack balance."

[**Example 2-10**]   Assuming (AX) =2107 H, analyze the following instruction functions and stack changes.

```
PUSH AX
POP  BX
```

The change in the stack before and after the execution of the instruction "PUSH AX" is shown in Figure 2-32.

The stack changes before and after the execution of the instruction "POP BX" are shown in Figure 2-33.

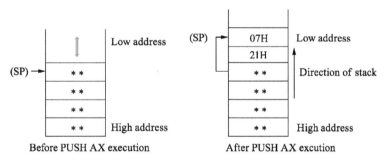

Figure 2-32   Execution instruction "PUSH AX" stack change

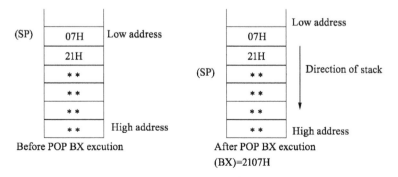

Figure 2-33   Execute the instruction "POP BX" stack change

[**Example 2-11**] Determine the order of right and wrong, and pointed out the cause of the error.

```
PUSH 1234H      ; faulty, PUSH instruction does not support immediate
                  addressing
POP  CS         ; wrong, CS can not be used as the destination operand
                  for the POP instruction
PUSH BL         ; Error, operand must be word type
```

(3) Exchange instructions XCHG

The exchange of data between general-purpose registers and general-purpose registers or storage units can be easily performed by using switch instructions.

◇ Instruction format and function:

```
XCHG  dst,src        ; (dst)↔(src)
```

◇ Supported addressing modes:

```
src:reg、mem
dst:reg、mem
```

The exchange instruction XCHG exchanges the contents of the source operand src with the contents of the destination operand dst. The direction of the exchange instruction is shown in Figure 2-34.

E. g:

```
XCHG AL,AH
XCHG SI,BX
XCHG [SI+3],AL
XCHG [DI+BP+3],BX
```

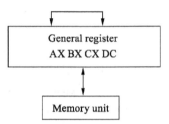

Figure 2-34 The direction of the exchange instruction

◇ XCHG instruction use note:

① src, dst can not be segment register seg and immediate imm.
② src, dst can not simultaneously for the memory operand mem.
③ src, dst must have the same type, that is, with the same byte type or the same type.

2. Accumulator dedicated transfer instructions

The accumulator (AX) dedicated transfer instruction includes the I/O data transfer instruction IN/OUT and the byte conversion instruction XLAT. The I/O data transfer instruction is used to complete the data transfer between the accumulator AL/AX and the I/O port. The byte conversion instruction XLAT is used to complete a byte conversion in memory. The input command IN and the output instruction OUT are described in Chapter 6.

Byte conversion instruction XLAT (also known as escape command or lookup instruction):

(1) Instruction format and function:

```
XLAT ; (AL)←((BX)+(AL))
```

This is an implied operand instruction, the implied operand is AL. The XLAT instruction sends the contents of a byte in the corresponding memory cell with the valid address of EA = (BX)+(AL) to the transcoding of one byte in AL, that is, (AL) ← ((BX)+(AL)).

The specific steps for implementing transcoding with XLAT are as follows:

① Create a transcoding table with a maximum capacity of 256B, locate the table in a contiguous address of a logical segment in memory, and place the valid address of the first address in the BX. In this way, BX points to the first address of the table.

② The data to be converted in the table number (index value) into the AL, the serial number is actually a table in the table and the first address between the amount of displacement.

③ Execute XLAT.

(2) XLAT instruction to use note:

① The index value is stored in the 8-bit AL register, so the length of the byte table can not be created more than 256.

② The number (index value) of the element in the table is counted from zero.

③ XLAT instruction default conversion table in the data segment DS, but can be carried out beyond the paragraph. In this case, you need to use another format of the byte conversion instruction:

```
XLAT    table
```

Where the operand table is the variable name of the byte table, that is, the first address of the table. In this way, the variable name table plus paragraph beyond the prefix can be achieved beyond the paragraph. This format is used only for the segment to go beyond or to improve the readability of the program. The first address of the byte table should be placed in front of BX.

④ The instruction does not affect the flag.

[**Example 2-12**]   Memory data segment has a hexadecimal ASCII table, as shown in Figure 2-35, the first address for the Hex-table, it is hoped that through the look-up table to get A ASCII value, and the results into the AL.

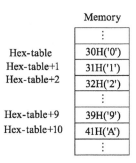

Figure 2-35   Diagram of Example 2-12

According to the meaning of the following can be prepared:

```
MOV BX, OFFSET Hex-table      ; BX←Header address
MOV AL, 0AH                   ; AL←Number of the elements to be
                                checked in the table
XLAT Hex-table                ; look-up table conversion
```

After the above instructions are executed in sequence, the ASCII code of 'A' is stored in AL, ie (AL) = 41H.

3. Address transfer instructions

Address transfer instructions include LEA, LDS, LES. This is a class of instructions that send an address code that can transfer the operand's segment address or valid address to the specified register.

(1) Effective address loading instruction LEA (Load Effective Address)

◇ Instruction format and function:

```
LEA dst, src      ; (dst)←src effective address EA
```

◇ Supported addressing modes:

```
src: mem
dst: reg16
```

The LEA instruction gets the 16-bit valid address EA of the source operand src (which must be a memory operand) and is passed to the 16-bit general register reg16 specified by dst.

[**Example 2-13**] LEA BX, [SI +0500H], assuming (DS) =3000H, (SI) =0100H.

The execution of the above instructions is shown in Figure 2-36. The result is (BX) = 0600H. Note that BX is the offset address 0600H, not the contents of the storage unit 1234H.

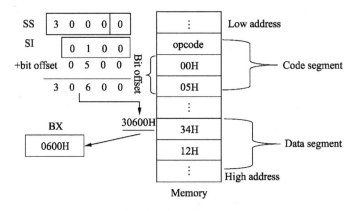

Figure 2-36 LEA instruction execution

◇ Note the difference between the LEA instruction and the MOV instruction:

The LEA instruction sends the valid address of the memory cell to the specified register, and the MOV instruction transfers the operand in the memory cell.

If the instruction in Example 2 is changed to MOV BX, [SI +0500H].

The result of the execution of the instruction is (BX) =1234H.

(2) Address Pointer Load DS instruction LDS (Load Pointer Using DS)

◇ Instruction format and function:

```
LDS  dst,src      ; (dst)←(src)
                  ; (DS)←(src +2)
```

◇ Supported addressing modes:

```
src: mem
dst: reg16
```

The LDS instruction is a command that conveys a 32-bit address pointer, starting with the memory cell specified by the instruction source operand src, reading four consecutive memory cell contents, a 32-bit address pointer, the first two bytes enter the register specified by dst in the instruction, and the last two bytes into the data segment register DS.

[**Example 2-14**]   LDS SI, [0010H].

Assume that the original (DS) = 0F000H, the contents of the memory cell are (0F0010H) =60H, (0F0011H) =01H, (0F0012H) =00H, (0F0013H) =20H.

The instruction execution process is shown in Figure 2-37, after instruction execution, (SI) =0160H, (DS) =2000H.

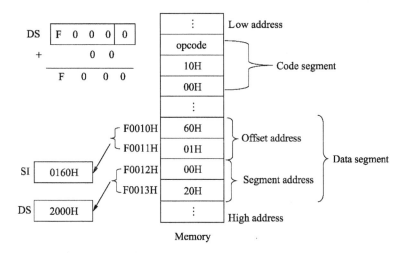

Figure 2-37   **LDS instruction execution**

(3) Address pointer load ES instruction LES (Load Pointer Using ES)
◇ Instruction format and function:

```
LES dst, src         ; (dst)←(src)
                     ; (ES)←(src+2)
```

◇ Supported addressing modes:
```
src: mem
dst: reg16
```

The LES instruction is similar to the LDS instruction and is a command to transfer a 32-bit address pointer. The function starts with the memory cell specified by the instruction source operand src, reads 4 consecutive memory cell contents, that is, a 32-bit address pointer, The first two bytes are sent into the register specified by dst in the instruction, and the last two bytes are sent to the additional segment register ES.

◇ Address transfer instruction LEA, LDS, LES use Note:
① dst can not be a segment register.
② src must use the memory addressing mode.
③ Does not affect the flag.

4. Flag register transfer instruction

The flag register transfer instruction includes LAHF, SAHF, PUSHF, POPF, which are used to complete the operation related to the flag. The operands in the instruction are specified in an implicit way, and the implied operands are AH and FLAGS registers, respectively.

(1) Take the sign instruction LAHF (Load AH from Flags)

◇ Instruction format and function:

    LAHF; (AH)←(Flags low byte)

The LAHF instruction transfers the lower 8 bits of the flag register FLAGS to the AH register. The execution of the LAHF instruction does not affect the flag bit.

(2) Flag flag SAHF (Stroe AH into Flags)

◇ Instruction format and function:

    SAHF; (Flags low byte)←(AH)

The transmission direction of the SAHF instruction is opposite to the LAHF instruction and the contents of the AH register are transferred to the lower 8 bits of the flag register FLAGS.

SAFF instruction execution flag, SF, ZF, AF, PF and CF in the FLAGS register will be modified to the value corresponding to the AH register, but the flags in the upper 8 bits of the FLAGS register are not affected, ie OF, DF, IF And TF.

(3) Sign push instruction PUSHF (Push Flags stack Stack)

◇ Instruction format and function:

    PUSHF ; (SP)←(SP) -2,
    ((SP))←(FLAGS low byte),((SP)+1)←(FLAGS high byte)

The PUSHF instruction performs a push operation, pushing the contents of the flag register FLAGS (16 bits) onto the stack, and the instruction execution does not affect the flag bit.

(4) Mark the stack instruction POPF (Pop Flags off Stack)

◇ Instruction format and function:

    POPL; (FLAGS low byte)←((SP)), (FLAGS high byte)←((SP)+1)
    ; (SP)←(SP)+2,

The POPF instruction, in contrast to the PUSHF instruction, performs a stack operation to pop the contents of the current stack to the flag register. The implementation of the POPF instruction has an effect on the flag.

In the 8086/8088 instruction system, there are corresponding setting and reset instructions for the OF, DF, and IF in the upper 8 bits of the flag register FLAGS, but there is no corresponding operation instruction for TF, and the PUSHF and POPF instructions can be used for this to achieve the TF value of the changes.

[**Example 2-15**]   The TF flag is cleared.

```
PUSHF           ; Push the contents of the FLAGS register onto the stack
POP AX          ;the stack operation sends the contents of the FLAGS register
                 to the accumulator AX
AND AH, 0FEH    ;Clear the lowest bit of the AH (corresponding to the TF bit)
PUSH AX         ;Push the contents of the AX onto the stack
POPF            ;flag pop-up stack, to achieve the TF value of the changes
```

5. Symbol expansion instructions

The symbol extension instruction, also called the type conversion instruction, includes the byte extension instruction CBW and the word extension instruction CWD. Its operation does not affect the flag.

When performing various arithmetic operations, the word length of the two operands in the instruction must meet the corresponding requirements. Specifically, in the addition, subtraction, and multiplication instructions, the word length of the two operands must be equal. In the division instruction, the dividend must be a double word length of the divisor. Therefore, in some cases it is necessary to expand the 8-digit to 16-bit or 16-bit to 32-bit. For the unsigned number, the extended word length only need to add enough zero in the upper part, and for the signed number, the extension word length should be added in the upper part of the corresponding sign bit, that is, the high part of the positive zero , the negative part of the high number of Tim 1. Symbol expansion. Although the number of data bits get longer, the data size has not changed.

(1) Byte extended instruction CBW (Convert Byte to Word)

◇ Instruction format and function:

```
CBW ; (AX)←(byte in the AL sign extended word)
```

This is an implied operand instruction, with implied operands AL and AH.

The CBW instruction extends the most significant bit D7 (AL symbol bit) of AL to AH to implement a byte shift from byte type to word type. That is, if AL is D7 =0, then (AH) = 00H; if AL is D7 =1, then (AH) =0FFH. The contents of AL are unchanged.

(2) word expansion instruction CWD (Convert Word to Double Word)

◇ Instruction format and function:

```
CWD ; (DX, AX)←(Word characters in AX are expanded into double words)
```

This is an implied operand instruction with implied operands of AX and DX.

The CWD instruction extends the most significant bit D15 (AX sign bit) of AX to DX to implement a conversion from word type to double word type. That is, if AX's D15 =0, then (DX) =0000H; if AX's D15 =1, then (DX) =0FFFFH. AX content unchanged.

[**Example 2-16**] Assume (AX) =0BA45H.

```
CWD       ; (DX) =0FFFFH,(AX) =0BA45H
CBW       ;(AX) =0045H
```

2.7.2 Arithmetic operations instructions

The 8086/8088 arithmetic operation instructions can handle four types of numbers, namely, unsigned binary numbers, signed binary numbers (complement representation), unsigned compressed decimal numbers (compressed BCD code), and unsigned non Compressed decimal number (uncompressed BCD code). Binary numbers can be 8 bits or 16 bits. In addition to compression decimal number can only plus, subtraction operation, the other three kinds can be added, subtracted, multiplied, in addition to operations.

This type of instruction will affect the status flag based on the results of the operation, and some also need to use some signs, use them when you pay attention to the status flag.

1. Addition and subtraction instruction

The addition instructions include ADD, ADC, INC. Subtracting instructions include SUB, SBB, DEC, NEG, CMP. In addition to the INC and DEC commands, the addition and subtraction instructions affect the six status flags CF, PF, AF, ZF, SF and OF in the flag register FLAGS. INC, DEC instructions do not affect the CF flag, but affect the remaining five status flags.

(AND, OR, XOR, TEST) in the logical operation to be described later is the same as that of the double operand instruction (ADD, ADC, SUB, SBB, CMP) in the addition and subtraction operation. Described as follows:

◇ Universal format:

```
OP dst,src
```

◇ Supported addressing modes:

```
src: reg,mem,imm
dst: reg,mem
```

◇ Addition and subtraction operation dual operation instruction use note:
① OP refers to a command mnemonic, not the actual instruction.
② src, dst can not simultaneously for the memory operand mem.
③ src, dst not for the segment register seg.
④ dst can not be immediate.
⑤ src, dst must have the same type, that is, with the same byte type or the same type.

(1) Addition instruction ADD (Binary Addition)

```
ADD dst, src; (dst)←(dst) + (src)
```

The ADD instruction adds the destination operand to the source operand and returns the result to the destination operand.

(2) Subtraction instruction SUB (Binary Subtraction)

```
SUB dst, src; (dst)←(dst) - (src)
```

The SUB instruction subtracts the source operand from the destination operand and returns the result to the destination operand.

(3) With carry add instructions ADC (Addition with Carry)

```
ADC dst, src; (dst)←(dst) + (src) + (CF)
```

The ADC instruction adds the destination operand to the source operand, plus the value of the carry flag CF, and returns the result to the destination operand.

(4) With a borrow subtraction instruction SBB (Subtraction with Borrow)

```
SBB dst, src; (dst)←(dst) - (src) - (CF)
```

The SBB instruction subtracts the source operand from the destination operand, subtracts the value of the borrow flag CF, and returns the result to the destination operand.

ADC, SBB instructions are mainly used to combine the ADD and SUB instructions to achieve multi-precision addition and subtraction operations.

CHAPTER 2　ASSEMBLY LANGUAGE BASICS

(5) Comparison instruction CMP (Compare)

```
CMP dst, src; (dst) - (src)
```

The CMP instruction subtracts the source operand from the destination operand, but the result is not sent back to the destination operand. After the CMP instruction is executed, the contents of both operands are unchanged, and the comparison result is only reflected in the flag, which is the difference between the CMP instruction and the subtraction instruction SUB. The CMP instruction is used to compare the size of two operands by subtracting the influence flag. Often used in conjunction with conditional transfer instructions to complete various conditional judgments and corresponding program transfers

(6) Seeking instructions NEG (Negate)

```
NEG dst; (dst)←0 - (dst)
```

The operation of the NEG instruction is to subtract the destination operand from "0" and return the destination operand.

(7) Plus 1 instruction INC (Increment by 1)

```
INC dst; (dst)←(dst) +1
```

The INC instruction increments the destination operand by 1 and returns the destination operand.

(8) Decrement 1 instruction DEC (Increment by 1)

```
DEC dst; (dst)←(dst) -1
```

The DEC instruction decrements the destination operand by 1 and returns the destination operand.

◇ INC, DEC, NEG instructions Note:

① Single operand instructions, operand dst can be register reg or memory operand mem, but can not be immediate imm and segment register seg.

② The operand of the instruction can be a byte or word type. If the operand is a memory operand mem, notice the explicit type, otherwise it is an error instruction.

E. g:

```
INC AL              ; 8-bit register plus 1
DEC BX              ; 16-bit register minus 1
DEC BYTE PTR [DI]   ; The memory operand is decremented by 1, byte operation
NEG WORD PTR [SI]   ; memory operand complement, word operation
```

③ INC, DEC instructions do not affect the CF logo, but affect the SF, ZF, AF, PF and OF signs.

④ The NIG instruction affects the six status flags CF, PF, AF, ZF, SF and OF in the flag register FLAGS.

[**Example 2-17**]　Test the following instructions to execute the results and the status of the flag.

```
MOV BX, 0       ; (BX) = 0, does not affect the flag
DEC BX          ; (BX) = 0FFFFH, CF does not affect, PF = 1, AF = 1, ZF = 0,
```

```
                            SF = 1, OF = 0
    INC BX                  ;(BX) = 0, CF does not affect, PF = 1, AF = 1, ZF = 1, SF = 0,
                            OF = 0
    SUB BX, 1               ;(BX) = 0FFFFH, CF = 1, PF = 1, AF = 1, ZF = 0, SF = 1, OF = 0
    NEG BX                  ;(BX) = 1, CF = 1, PF = 0, AF = 1, ZF = 0, SF = 0, OF = 0
```

Readers can use DEBUG debug the sequence of instructions to observe the implementation of the note, after the completion of each instruction after the completion of BX and the flag changes.

[**Example 2-18**]   The data segment is defined as follows, try to write the instruction sequence to calculate the sum of the variables DVAR1 and DVAR2 and store the result in the SUM variable.

```
    DSEG SEGMENT            ; data segment
      DVAR1 DD 12345678H
      DVAR2 DD 89ABCDEFH
      SUM DD?
    DSEG ENDS
```

Solution: two variables are double word variable, because the 8086/8088 can only handle up to 16-bit addition, so the addition should be carried out in two, first low 16-bit add, and then do the high 16-bit add, The addition of the upper 16 bits must take into account the carry of the lower 16 bits. It can be achieved with the following sequence of instructions.

```
    MOV AX, WORD PTR DVAR1       ; DVAR1 low 16 bits to send the accumulator AX
    ADD AX, WORD PTR DVAR2       ; DVAR1 is added to the lower 16 bits of DVAR2
    MOV WORD PTR SUM, AX         ; low 16 bits and send SUM low 16 bits save
    MOV AX, WORD PTR DVAR1 + 2   ; DVAR1 high 16 bits to send the accumulator AX
    ADC AX, WORD PTR DVAR2 + 2   ; DVAR1 is added to the upper 16 bits of DVAR2
    MOV WORD PTR SUM + 2, AX     ; high 16 bits and send SUM high 16 bits save
```

In the above block, two different addition instructions are used for ADD and ADC. The ADD instruction is used to complete the addition of two bytes of the lower 16 bits. The result of the addition may generate a carry, so the upper 16 bits sum must take into account the status of the carry flag.

2. Multiplication

8086/8088 system multiplication instructions have two: MUL, IMUL, which can be completed 8-bit or 16-bit binary multiplication.

(1) Unsigned multiplication instruction MUL (Multiplication Unsigned)

```
    MUL src                  ; when src is the byte amount, the byte multiplication
                               is performed: (AX)←(AL) × (src)
                             ; When src is a word, the word is multiplied: (DX,AX)←
                               (AX) × (src)
```

(2) Symbolic multiplication instruction IMUL (Integer Multiplication)

```
    IMUL src                 ; When the byte is multiplied by the number of bytes:
                               (AX)←(AL) × (src)
```

CHAPTER 2　ASSEMBLY LANGUAGE BASICS

;When the src is a word: (DX, AX)←(AX)×(src)

◇ Note　when use MUL, IMUL instruction:

① src can be register reg or storage unit mem, can not be immediate and segment register seg.

② MUL, IMUL is a single operand instruction, and the other multiplier involved in the multiplication is implied as the accumulator AL (byte multiplication) or AX (word), the product implicitly stores AX (byte multiplication) or DX and AX (word multiplication, where the product is 16 bits in the DX and the lower 16 bits are stored in the AX). The operand of the multiplication and the result of the operation are shown in Figure 2-38.

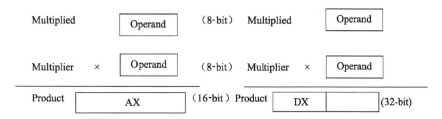

Figure 2-38　Operand of the multiplication and the result of the operation

③ src if the storage unit mem, pay attention to clear type.

E. g:

```
MUL BYTE PTR [SI + disp]      ; AL multiplied by 8-bit memory operand,
                                multiplied by AX
MUL WORD PTR [BP][DI]         ; AX multiplied by 16-bit memory operand,
                                multiplied by DX: AX
```

④ Multiplication instructions according to the following rules affect the CF and OF signs.

MUL instruction-if the higher half of the product (AH or DX) is 0, then OF = CF = 0; otherwise OF = CF = 1.

IMUL instruction-if the higher half of the product is half the sign extension, then OF = CF = 0; otherwise OF = CF = 1.

The multiplication instruction uses OF and CF to determine whether the higher half of the product has a valid value, rather than whether an overflow occurs or a carry occurs.

⑤ The multiplication instruction is not defined for the PF, AF, ZF, SF flags.

◇ **Special attention**: "No definition of the mark" means that the signs are arbitrary, unpredictable (ie, do not know 0 or 1) after the instruction is executed. This is different from the "sign does not affect" which does not affect the implementation of the instructions do not change the original state of the logo.

[**Example 2-19**]　If (AL) = 0FFH, (BL) = 2, respectively use MUL and IMUL to calculate (AL)×(BL).

```
MUL BL                        ; product (AX) = 01FEH, (255 ×2 = 510)
IMUL BL                       ; product (AX) = 0FFFEH, (-1 ×2 = -2)
```

3. Division instructions

8086/8088 CPU implementation of the division operation is specified:

① Divisor can only be half the length of the dividend, that is, when the dividend is 16, the divisor should be 8, the divisor is 32, the divisor should be 16.

② When the dividend is 16 bits, it is should be stored in the AX, 8-bit divisor can be stored in the register or memory, the division of the results of the 8-bit stored in the AL, and 8-bit remainder stored in the AH.

③ When the dividend is 32 bits, it should be stored in the register pair consisting of DX and AX (the upper 16 bits in the DX, the lower 16 bits in the AX), the 16-bit divisor can be stored in the register or memory, the division result of the 16-bit business stored in the AX, and 16-bit remainder stored in the DX.

The operand of the division operation and the result of the operation are shown in Figure 2-39.

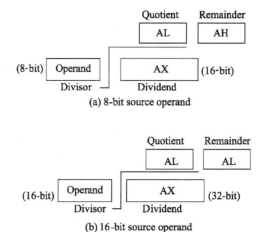

Figure 2-39 Operand and operation result of the division operation

8086/8088 division instructions have two: DIV, IDIV, for the single operand instruction. Only the specified divisor is displayed in the division instruction, and the rest is an implied operand.

(1) Unsigned number division instruction DIV (Division Unsigned)

```
DIV src
(AX)←(AX)/(src), (AH)←(AX)%(src) When the src is the byte amount,
(DX:AX)/(src), (DX)←(DX:AX)%(src) When the src is the word.
```

(2) Signed number division instruction IDIV (Integer Division)

```
IDIV src
(AX)←(AX)/(src), (AH)←(AX)%(src) When the src is the byte amount,
(DX:AX)/(src), (DX)←(DX:AX)%(src) When the src is the word.
```

◇ DIV, IDIV instruction use Note:

① Divisor src can be register reg or storage unit mem, can not be immediate and segment register seg.

② Divisor src if the storage unit mem, pay attention to clear type.

E. g:

```
IDIV BX ; DX              ; AX divided by BX
IDIV BYTE PTR [DI]        ; AX divided by 8-bit memory operand
IDIV WORD PTR [BX][SI]    ; DX: AX divided by 16-bit memory operand
```

③ The instruction has no definition of the 6 status flags in the flag register FLAGS.

④ When the divisor is 0 or the quotient is out of the AL/AX representation range (division overflow), the CPU generates an internal interrupt of type 0, that is, the division error is interrupted.

⑤ Do not allow two words equal to the number of equal division. If the dividend is equal to the divisor, the divisor should be type converted before the division so that it is twice as long as the divisor.

Special attention: IDIV instructions should be used before the CBW, CWD instruction is to convert, but in the absence of symbols in addition to DIV before, you should not use CBW or CWD instructions, generally use XOR and other instructions to clear 8 or 16 high.

⑥ The sign of the remainder of the IDIV instruction is always the same as the sign of the dividend.

[**Example 2-20**]   Calculate (AX)/(BL) with DIV and I DIV. Known (AX) = 0410H, (BL) =0B8H.

```
DIV BL    ; Quotient AL = 05H, remainder AH = 78H
IDIV BL   ; Quotient AL = F2H (i.e., -14), the remainder AH = 20H (i.e., 32)
```

The symbol number 0410H true value of 1040, B8H true value of -72. Note the difference between DIV and I DIV. Readers can use DEBUG debug verification analysis.

[**Example 2-21**]   X, Y, Z, V, and W are 16-bit signed numbers, calculate W ← (V − (X ∗ Y + Z −1234))/X.

Analysis: assembly language, for the expression of the calculation, must be in accordance with the level of the operator at all levels, reasonable arrangements for the order and data types. Therefore, we must first determine the order of calculation, followed by the data to determine the type of step. For the expression to be calculated in this example, the order of calculation is as follows:

① X ∗ Y→temporary intermediate results. 16-bit signed number multiplied, the product is 32 bits.

② X ∗ Y + Z→X ∗ Y + Z − 1234 → Storing intermediate results. Because X ∗ Y results for 32-bit, so Z should be extended to 32-bit symbols, 1234 for the immediate, positive, direct 0 can be extended. So this step is a 32-bit addition and subtraction.

③ V − (X ∗ Y + Z − 1234) → (V − (X ∗ Y + Z − 1234))/X → Save the final result. Since the result of X ∗ Y + Z-1234 is 32 bits, the V requires the sign to be expanded to 32 bits after the operation. (X ∗ Y + Z − 1234)) is 32 bits, X is 16 bits, so (V − (X ∗ Y + Z-1234))/X bit 16 bits, the final result is 16 bits.

The block is as follows:

```
MOV AX, X
```

```
        IMUL Y              ; X * Y
        MOV CX, AX          ; X * Y (32 bits) Temporary access to BX; CX. DX; AX is
                              required for symbol expansion of Z
        MOV BX, DX
        MOV AX, Z           ; Z Insert AX for symbol expansion to DX; AX, extended to 32
                              bits
        CWD
        ADD CX, AX
        ADC BX, DX          ; X * Y + Z
        SUB CX, 1234
        SBB BX, 0           ; X * Y + Z -1234
        MOV AX, V           ; V Insert AX for symbol expansion to DX; AX, extended to 32
                              bits
        CWD
        SUB AX, CX
        SBB DX, BX          ; V-( X * Y + Z -1234 )
        IDIV X              ; (V-( X * Y + Z -1234 )) /X, in AX, the remainder is in DX.
        MOV W, AX           ; save the final result
```

4. Decimal adjustment instructions

The computer arithmetic operations are for the binary number of operations, and decimal system is used in daily life. In the 8086/8088 system, there are a class of decimal adjustment instructions for decimal arithmetic operations.

In the computer system with BCD code that decimal number, BCD code has two ways: one for the compressed BCD code, that is, each byte that has the two BCD number; the other is called the non-compressed BCD code, that is, One byte represents a BCD number, where the upper four bits are filled with 0. For example, the decimal number 1234D, expressed as the compressed BCD number is: 1234H; expressed as uncompressed BCD number: 01020304H, expressed in 4 bytes. The relevant BCD conversion instructions are shown in Table 2-12.

Table 2-12 Decimal Adjustment Instructions

Instruction format	Instructions for instructions
DAA	Compressed BCD code addition adjustment
DAS	Compressed BCD code subtraction adjustment
AAA	Uncompressed BCD code addition adjustment
AAS	Uncompressed BCD code subtraction adjustment
AAM	Multiplication of BCD after adjustment
AAD	Division before the BCD code adjustment

Note: When the BCD code is multiplied and multiplied, the unsigned number is used, so that AAM and AAD should appear before MUL and before DIV.

[**Example 2-22**] Assuming AL is the BCD code of 28, BL is the BCD code of 68, and find the sum of these two decimal numbers. The sequence of instructions is as follows:

```
        ADD AL, BL
```

DAA

(AL) =28H, (BL) =68H, after the implementation of ADD: AL =90H, AF =1; and then the implementation of DAA instructions, the correct results: AL =96H, CF =0, AF =1. DAA is the correct result after adjustment (28 +68 =96).

2.7.3 Logical and shift class instructions

The logical and shift class instructions can logically or shift operations in bits or bits in 8-bit or 16-bit registers or memory locations. This type of instruction includes logic operations instructions, shift instructions, and cyclic shift instructions 3 group.

This type of instruction also affects the status flags, and pay attention to the status flags when using them.

1. Logic instruction

Logical instructions include NOT, AND, OR, XOR, and TEST.

NOT is a single operand instruction. The supported operand is the same as the INC and DEC instructions, but the NOT instruction does not affect the flag.

The AND, OR, XOR, and TEST instructions are dual operand instructions. The supported operand form is the same as the double operand instruction (for example, ADD, SUB, etc.) of the addition and subtraction class. SF, PF, ZF are set according to the characteristics of the operation result; AF is not defined and the status is uncertain. The operation of the four logical operation instructions is affected by the flag.

(1) Logical "NOT" instruction NOT (Logical not)

```
NOT dst
```

The NOT instruction inverts the 8-bit or 16-bit operand bit by bit and returns the destination operand.

(2) Logical "AND" instruction AND (Logical and)

```
AND dst, src; (dst)←(dst)∧(src)
```

The AND instruction causes the logical AND operation of the destination operand and the source operand bit by bit, and the result is returned to the destination operand.

(3) Logical OR command OR (Logical Inclusive or)

```
OR dst, src; (dst)←(dst)∨(src)
```

The OR instruction causes the logical OR operation of the destination operand and the source operand bit by bit, and the result is returned to the destination operand.

(4) Logical "exclusive OR" instruction XOR (Logical Exclusive or)

```
XOR dst, src; (dst)←(dst)⊕(src)
```

The XOR instruction causes the logical AND divisor of the destination operand and the source operand bit by bit, and returns the destination operand.

(5) Test instruction TEST (Test or non-Destructive Logical and)

```
TEST dst, src; (dst)∧(src)
```

The operation of the TEST instruction is similar to the AND instruction, which means that the destination operand and the source operand are logically ANDed, and the difference between the two is that the TEST instruction does not return the result of the logical operation to the destination operand, and the result of the logical operation is only reflected in the status flag.

The AND instruction is typically used to mask and hold bits, where the bits to be masked can be ANDed with "0", and the bits to be retained can be ANDed with "1".

[**Example 2-23**] AX in the most significant and the lowest bit to retain the remaining bits cleared, the following instructions can be used:

```
AND AX, 8001H
```

◇ OR instruction is often used to set some bits while keeping the remaining bits unchanged. The bits that need to be set are logically OR with "1", while the bits that remain unchanged are logically OR with "0".

[**Example 2-24**] The lower 4 bits of BX are set, and the remaining bits are unchanged, and the following instruction can be used:

```
OR BX, 000FH
```

◇ ANDAND, OR instruction has a common characteristic: if a register operand itself is logically "AND" or logical "or", its contents remain unchanged, but the logic operation itself changes the state of the flag bit. Said that will affect SF, ZF and PF, and make OF and CF clear. With this feature, it is possible to determine whether it is the positive or negative data, whether it is zero or not, and the parity characteristics, by logical operation, after the data transfer instruction.

E. g:

```
MOV AL, BVAR
AND AL, AL       ; Impact flag
JNZ NEXT         ; if not zero, move to NEXT
...
NEXT:...
```

In the above procedure, if there is no logical operation, the conditional judgment and program transfer can not be performed after the MOV instruction. Since the MOV instruction does not affect the flag bit state, it is of course possible to use other instructions instead of logical instructions such as CMP AL, 0 Or SUB AL, 0, etc., but relatively speaking, the less the logic of instruction bytes are, the faster the implementation is.

◇ XOR command is often used to some specific bit "inversion", while the remaining bits remain unchanged, which to "seek" the bit and "1" logic "XOR", to keep the same bit and "0" for logical exclusive OR.

[**Example 2-25**] Assumptions (BH) = 10110010B, analyze the contents of the following instructions after the implementation of BH.

```
XOR BH, 01011011B
```

After instruction execution, (BH) =11101001B.

◇ Another important application of the XOR instruction is that a register operand itself is logically "exclusive-OR" to itself, for example:

```
XOR BH, BH      ; BH cleared
XOR SI, SI      ; SI is cleared
```

Of course, the use of other instructions can also achieve the contents of the register clear, for example:

```
MOV SI, 0       ; SI clear
SUB SI, SI      ; SI is cleared
AND SI, 0       ; SI is cleared
```

◇ TEST instruction is often used for bit testing, and together with the conditional transfer instruction to complete the judgment of the specific bit, and to achieve the corresponding program transfer. This is similar to the comparison instruction CMP, but the TEST instruction only compares certain bits, and the CMP instruction compares the entire operand.

To detect whether the least significant bit in AL is 1, if it is a branch, use the following command:

```
TEST AL, 01H
JNZ NEXT
...
NEXT:
```

To detect whether the contents of BX are 0, if the transfer is 0, the following instructions are available:

```
TEST BX, 0FFFFH
JZ NEXT
...
NEXT:
```

2. Shift instructions

8086/8088 system, the shift instruction can be divided into two groups of acyclic shift instructions and cyclic shift instructions.

The non cyclical shift instruction has two types of logical shift and arithmetic shift. Logical shift is an unsigned number shift, and is always empty the bit with "0"; the arithmetic shift moves the signed number, and the sign bit must not change during the shift. There are four non-cyclic shift instructions: SHL, SHR, SAL, SAR (where the SHL and SAL instructions operate exactly the same), and the operations they perform are shown in Figure 2-40.

(1) Logical left shift SHL (Shift Logical Left).
(2) Arithmetic left shift SAL (Shift Arithmetic Left).
(3) Logical right shift SHR (Shift Logical Right).
(4) Arrow Right Shift (Shift Arithmetic Right).

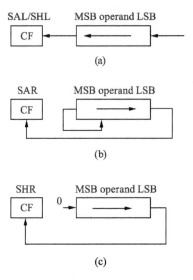

Figure 2-40 Functions of acyclic shift instructions

The cyclic shift instruction has two types of cyclic shift and carry cyclic shift. The so-called cyclic shift, refers to the shift object is connected to the end of the data, the data in the closed loop loop and not lost. The 8086/8088 system has four cyclic shift instructions: ROL, ROR, RCL, RCR. The operations they perform are shown in Figure 2-41.

(1) Rotation left without rotation ROL (Rotate Left).
(2) Rotation right without rotation ROR (Rotate Right).
(3) Rotation left hand with RCL (Rotate Left through Carry).
(4) Rotation Right Carry Carry RCR (Rotate Right through Carry).

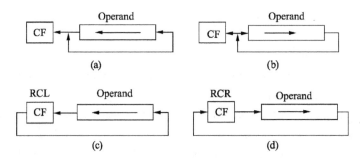

Figure 2-41 Function of the cyclic shift instruction

The above two sets of shift class instructions support the same form of operands. Uniform description is as follows:

◇ Universal format:

```
OP dst, src
```

◇ Supported addressing modes:

```
src: 1, CL
dst: reg, mem
```

◇ Shift class instructions use Note:

① OP refers to a command mnemonic, not the actual instruction.

② src is used to specify the number of shift bits, only 1 or CL.

③ dst is used to specify the destination operand to be used for the shift operation. It can be a register reg or a storage unit mem, which can not be a segment register seg and an immediate value.

④ Shift instructions on the flag register flags in the six state signs of the impact, as follows:

· CF = moved in value.

· When the shift bit is 1, if the value of the most significant bit changes, then OF = 1, otherwise OF = 0; When the number of shifts is greater than 1, OF is not determined.

· The acyclic shift instruction sets the SF, ZF, and PF flags according to the shift result, and is not defined for the AF flag.

· The cyclic shift instruction does not affect the SF, ZF, PF, AF flags.

⑤ dst can be byte or word type. If the storage unit mem, pay attention to clear type, otherwise the instruction will be wrong.

E. g:

```
SHR WORD PTR [BX + DISP], CL    ; word memory cell right shift CL
                                  indicates the bit
ROL BH, 1                       ; register cycle shifted by 1 bit
                                  left
ROR BYTE PTR [BP][DI], CL       ; byte storage unit loop right shift CL
                                  to indicate the bit
RCL BYTE PTR disp [SI], 1       ; byte memory unit with carry carry
                                  left 1 bit
RCL WORD PTR [SI + BP], CL      ; word memory unit with carry cycle
                                  left shift CL to specify bit
```

◇ An unsigned binary number is shifted by 1 bit, which is equivalent to the number multiplied by 2, so that the multiplication can be done with the left shift instruction, and the shift instruction is much faster than the execution of the multiplication instruction.

[**Example 2-26**] Test the following block function.

```
SAL AL,1        ; (AL)←(AL)×2
MOV BL,AL       ; (BL)←(AL)×2
SAL AL,1        ; (AL)←(AL)×4
SAL AL,1        ; (AL)←(AL)×8
ADD AL,BL       ; (AL)←(AL)×8+(AL)×2
```

The function of the block is: to achieve (AL)←(AL)×10.

◇ An unsigned binary number is shifted right by one bit, which is equivalent to the number divided by 2, so that the division can be done with the right shift instruction, and the shift instruction is much faster than the divide instruction execution.

[**Example 2-27**] Divide the 16-bit unsigned number in AX by 512.

Because $2^9 = 512$, so as long as the AX logic right to move 9 to achieve the above division.

The sequence of instructions are as follows:

```
    MOV CL, 9         ; CL←Number of shifts
    SHR AX, CL        ; AX logic right shift 9 bits
```

◇ The use of a carry cyclic shift instruction and shift instruction combination can achieve multi-precision shift.

[**Example 2-28**]   Move the DX and AX combined 32-bit operands to the left by 1 bit.

Note that the highest bit in the lower 16 bits must be shifted to the least significant bit in the lower 16 bits during the shift, so the lower 16 bits of the AX can be shifted left by one bit, and the upper 16 bits of the lower 16 bits are shifted into the CF, Then the DX in the high 16 with CF left shift 1 bit. The sequence of instructions is as follows:

```
    SAL AX, 1         ; AX left one bit, the highest position of AX moves into CF
    RCL DX, 1         ; DX with carry carry left 1 bit, CF into DX's lowest bit
```

2.7.4 Program control class instructions

The program control class instruction is used to control the flow of the program. In the 8086/8088 system, the instruction execution order is determined by the contents of the code segment register CS and the instruction pointer register IP. In general, the instructions in the program are executed sequentially, that is, CS does not change, and the IP register automatically modifies the memory unit that points to the next instruction. But in actual operation, the program will often be based on the state of the microprocessor and some constraints, no longer in the order of implementation, which requires the use of procedures in the program control class instructions, in order to achieve branch and loop. The program control class instruction changes the execution order of the program by changing the contents of CS and IP. 8086/8088 system provides a large number of program control instructions, according to the function can be divided into the following four categories:

(1) Unconditional transfer instructions and conditional transfer instructions.
(2) Cycle instructions.
(3) Procedure call and process return instruction.
(4) Soft interrupt instruction and interrupt return instruction.

In addition to interrupt instructions, the remaining instructions do not affect the flag bit state.

Since the program code can contain multiple code segments, it is necessary to modify the code segment register CS according to whether it needs to be modified. It can be divided into intra-segment transfer and inter-segment transfer. Intra-segmentation refers to the instruction that continues execution after the transfer is still in the same code segment, and only the instruction pointer IP is reset. Conditional branch instructions and loop instructions can only perform intra-segment transfers. Inter-segment transfer refers to instructions that continue to be executed after the transfer in another code segment, not only to reset the IP, but also to reset the code segment register CS. Soft interrupt instructions and interrupt return instructions are generally inter-segment transfers. The unconditional branch instruction and the procedure call and the return instruction can be either intra-segment transfer or inter-segment transfer. Intra-segmental transfers are also known as near metastases, and inter-segment transfers are also known as distant metastases.

For unconditional branch instructions and procedure call instructions, according to determine the transfer destination address, can also be divided into direct transfer and indirect transfer. If the address is given directly in the directive, it is called a direct transfer; if the destination address is given indirectly in the instruction, it is called indirect transfer. Assembly language, in the expression of direct transfer, is the use of the target address of the label. When an indirect transfer is expressed, a register name or a memory operand is used.

1. Unconditional transfer instruction

◇ Instruction format and function:

```
JMP opr           ; Modify the IP or CS: IP with the destination address
                    specified by opr.
```

◇ Supported addressing modes:

Intra-segment direct addressing, intra-segment indirect addressing, inter-segment direct addressing, inter-segment indirect addressing.

The unconditional branch instruction JMP enables the program to change the execution order without any prerequisites. CPU as long as the implementation of JMP instructions, you can make the program to the specified destination address, from the target address to start the implementation of instructions.

(1) Within the direct transfer

The use of direct transfer instructions in unconditional segments is as follows:

JMP label; intra-segment transfer, direct addressing, (IP)←(IP) + displacement

This instruction makes the program flow be transferred unconditionally to the label address. The jump range is within the range of 64KB of the current code segment, and the address displacement is represented by a 16-bit number. The instruction can also be expressed as:

JMP NEAR PTR label; (IP)←(IP) +16 bit displacement

If the range of transfers is between −128 and +127, the address offset is represented by an 8-digit number, and we call this transfer a short transfer. The instructions can be expressed as:

JMP SHORT label; (IP)←(IP) +8 bit displacement

In actual use, the assembler will automatically determine the NEAR attribute or SHORT attribute of the label according to the actual jump range. Therefore, the attribute operator "NEAR PTR" or "SHORT" can be omitted from the label, and the assembler is judged by the label.

The transfer of the transfer destination address using the difference between the target address and the current IP address is also referred as relative transfer.

(2) Indirect transfer within the segment

The format of the indirect transfer instruction in the unconditional segment is as follows:

```
JMP opr
```

This instruction makes the control be transferred unconditionally to the destination address given by the contents of the operand opr. Instead of giving a label in the branch instruction, an operand is given, which can be a 16-bit general register or word storage unit. The 16-bit operand content is the destination valid address (EA) into the IP.

E. g:

```
        JMP CX                      ; CX register contents send IP
        JMP WORD PTR [1234H]        ; word storage unit [1234H] to send IP
```

(3) Direct transfer between paragraphs

Inter-segment transfers are also known as distant transitions. The format of the direct transfer instruction between unconditional segments is as follows:

```
        JMP FAR PTR label
```

This instruction places the segment address and offset address of the destination address into CS and IP, respectively, so that the program flow is unconditionally transferred to the address corresponding to the label. The symbol "FAR PTR" before the label indicates to the assembler that this is a segment transfer. This transfer method, which directly contains the transfer destination address in the instruction, is also called absolute transfer.

E. g:

```
        JMP FAR PTR EXIT; EXIT is the label defined in another code segment
```

(4) Between the indirect transfer

The format of the indirect transfer instructions between unconditional segments is as follows:

```
        JMP opr
```

This instruction causes the control to be transferred unconditionally to the destination address given by the contents of the operand opr. The operand opr must be a double word storage unit.

E. g:

```
        JMP DWORD PTR [1234H]       ; low word contents of double word memory unit
                                      send IP
                                    ; Double word storage unit to send the
                                      contents of the high word CS
```

2. Conditional transfer instructions

The conditional branch instruction needs to determine whether the program has been transferred according to the specified conditions. Therefore, the execution of the conditional branch instruction first tests the specified condition. If the condition is satisfied, the program moves to the destination address to execute the program; if the condition is not satisfied, the program will execute the next instruction sequentially. Thus implementing the branch program. 8086/8088 has a wealth of conditional transfer instructions, the vast majority (except JCXZ instruction) is a certain bit, or the logical operation of the flag as a test condition. The general instruction format is:

```
        JCC label
```

The "CC" in the instruction mnemonic indicates the test condition. The instruction in the instruction indicates the destination address of the transfer, but unlike the JMP instruction, the JCC instruction only supports short transfers, ie the relative amount of the next instruction to the destination address of the conditional branch instruction must be between −128 and +

127 (IP)←(IP) +disp, disp is the relative displacement of 8 bits. The transfer of the condition is similar to that of the JMP instruction.

8086/8088CPU all the conditions of the transfer instruction are shown in Table 2-13, the table slash separated by the same instruction a number of mnemonic form, some instructions using multiple mnemonic form is only convenient to remember and use.

Table 2-13 Conditional transfer instructions

Classification	Mnemonic	Transfer condition	Description
(1) Determine the status of a single flag	JZ/JE	ZF =1	zero / equal, then transferred
	JNZ/JNE	ZF =0	Not zero / not equal, then transfer
	JS	SF =1	negative, then transferred
	JNS	SF =0	positive, then transferred
	JO	OF =1	Overflow, then transfer
	JNO	OF =0	Not overflow, then transfer
	JP/JPE	PF =1	The number of low byte '1' is even, then transferred
	JNP/JPO	PF =0	The number of low byte '1' is odd, then transferred
	JC	CF =1	There is a carry, then transfer
	JNC	CF =0	No carry, then transfer
(2) Compare two unsigned numbers	JB/JNAE/JC	CF =1	Below / no more than equal, then transfer
	JNB/JAE/JNC	CF =0	Not less than / above or equal to, then transfer
	JBE/JNA	(CF $\vee$ ZF) =1	Less than or equal to / not above, then transfer
	JNBE/JA	(CF $\vee$ ZF) =0	Not less than equal to / above, then transfer
(3) Compare two signed numbers	JL/JNGE	(SF $\forall$ OF) =1	Less than / no more than equal, then transfer
	JNL/JGE	(SF $\forall$ OF) =0	Not less than / greater than or equal to, then transfer
	JLE/JNG	((SF $\forall$ OF) $\vee$ ZF) =1	Less than or equal to / not greater than, then transfer
	JNLE/JG	((SF $\forall$ OF) $\vee$ ZF) =0	Not less than or equal to / greater than, then transfer
(4) Test whether the CX value is zero	JCXZ	CX = 0	CX content is 0, then transfer

◇ Use conditional branch instruction note that the instruction that affects the status of the flag bit should be executed first before the condition flag can be used to determine whether the program is transferred. So before the conditional transfer instruction, often CMP, TEST instructions and other arithmetic, logic operations and other instructions.

Jcc instruction does not affect the flag, but to use the flag, according to the use of different signs, divided into four kinds of situations, see Table 2-13.

The order relation of unsigned numbers is called Above, Equal and Below. The order relation of signed numbers is called greater than, equal, and less. Regardless of unsigned or signed number, whether the two numbers can be reflected by the ZF logo. When two unsigned numbers are subtracted, the CF bit (whether or not there is a borrow) reflects the level of two unsigned numbers, so the conditional branch instruction (such as JB and JAE, etc.) for unsigned comparison CF is to determine whether the conditions are true. But the carry flag CF can not reflect the size of two signed numbers. The size relationship of the signed number is to be combined with the SF and OF flags. Therefore, conditional branch instructions (such as JL and JGE) are used to check the flag SF and OF to determine whether the condition is satisfied.

The number of symbols between the unsigned numbers, the number of symbols between the conditions after the transfer instructions are very different. In the use of attention to distinguish them, it can not be confused.

For example: the following block is to achieve AX and BX in the two unsigned number of comparison, the larger number stored in the AX, the smaller number stored in the BX:

```
CMP AX, BX
JAE OK; unsigned number of size transfer
XCHG AX, BX
OK: ...
```

If the two numbers to be compared are signed, the JAE in the above block should use the JGE instruction instead.

[**Example 2-29**] If the data stored in AX is odd, set CL to 0, and if it is even, set CL to −1.

Analysis: to determine whether AX is odd or even, as long as the AX minimum bit D0 is "0" (even), or "1" (odd). Therefore, it is conceivable to influence the flag bit according to the difference of the D0 bits to form a determination condition, and the conditional branch instruction is used. How can D0 bits affect the flag bit? Which flag is used to determine (which conditional transfer instruction is selected)? There are a number of ways, readers can also consider looking at other solutions.

Method 1: use the test command to change the other bits except the least significant bit to 0, leaving the least significant bit unchanged. Judge this data is 0, AX is even; otherwise, for the odd. Determine whether the result of the operation is 0, use the ZF flag, that is, the JZ or JNZ instruction. The block is as follows:

```
TEST AX, 01H
JZ EVE
MOV CL, 0
JMP NEXT
EVE: MOV CL, 0FFH
NEXT: ...
```

Method 2: move the least significant bit with the shift instruction to the carry flag to determine that the carry flag is 0, AX is even; otherwise, it is odd, using JNC or JC. The block is as follows:

```
MOV BX, AX
SHR BX, 1
JNC EVE
MOV CL, 0
JMP NEXT
EVE: MOV CL, 0FFH
NEXT: …
```

Method 3: move the least significant bit with the shift instruction to the most significant bit (sign bit). If the symbol flag is 0, AX is even; otherwise, it is odd, using JNS or JS instruction, the block is as follows:

```
MOV BX, AX
ROR BX, 1
AND BX, BX        ; Note why use this AND instruction here, can it also be
                    replaced by other instructions?
JNS EVE
MOV CL, 0
JMP NEXT
EVE: MOV CL, 0FFH
NEXT: …
```

[**Example 2-30**] DVAR1 and DVAR2 are defined signed double word variables. If DVAR1 > DVAR2, set the CL value to 1, otherwise set the CL value to −1.

Analysis: because DVAR1, DVAR2 for the double word variable, the 8086/8088 system can not directly compare the size, they need to be divided into high and low 16-bit comparison. If the high 16 bits are not equal, then their size relationship is determined, if equal, then need to further lower 16 bits. The block is as follows:

```
MOV DX, WORD PTR DVAR1+2
MOV AX, WORD PTR DVAR1
CMP DX, WORD PTR DVAR2+2
JG BIG
JL NOTBIG
CMP AX, WORD PTR DVAR2
JA BIG
NOTBIG: MOV CL, 0FFH
JMP NEXT
BIG: MOV CL, 01H
NEXT: …
```

In particular, in this case, the lower 16 bits are used with the JA instruction for the unsigned number, and the JG instruction can not be used (why? The reader is thinking).

3. Cyclic instructions

In the program design, often need to make some of the repeated implementation of the program, these repeated the implementation of the program is the cycle. The loop instruction is used to implement the loop control, which is actually a set of enhanced conditional branch instructions, and controls the transfer according to whether the test flag bit condition satisfies the condition. The 8086/8088 system has three loop control instructions: LOOP, LOOPZ/LOOPE, LOOPNZ/LOOPNE, which implicitly use the CX register as the cycle count counter. As shown in Table 2-14.

Table 2-14 Cycles Instructions Table

Instruction format	Perform the operation
LOOP Label	CX = CX −1; If CX≠0, then the label, the cycle
LOOPNZ/LOOPNE Label	CX = CX −1, If CX≠0 and ZF =0, then the rotation number, the cycle
LOOPZ/LOOPE Label	CX = CX −1, If CX≠0 and ZF =1, then the label, the cycle

◇ Use of cyclic instructions Note:

① The cyclic instruction has no effect on the flag.

② The cyclic instruction is the same as the conditional branch instruction, and only the short transfer is supported.

③ When using cyclic instructions, the number of loops should be sent to the CX register before the cycle routine starts.

④ LOOP instruction function is equivalent to the following two combinations of instructions, namely:

```
DEC CX
JNZ label
```

[**Example 2-31**] Programming calculation $1 +2 +3 + \cdots +100 = ?$ The result is stored in the word variable SUM.

The block is as follows:

```
XOR AX, AX           ; Accumulator is cleared
MOV BX, 0001H        ; BX ← 1
MOV CX, 100          ; CX←Number of cycles 100
AGAIN: ADD AX, BX    ; AX←(AX) +(BX)
INC BX               ; BX←(BX) +1
LOOP AGAIN           ; continue without looping
MOV SUM, AX          ; loop end, save the result
```

[**Example 2-32**] In the data segment, the STRING1 and STRING2 addresses 100 characters, compare the two strings, and find out that the first different characters are sent to the AL and BL registers respectively. If the two strings are identical, Then AL =BL =0.

```
LEA SI, STRING1       ; SI ← string STRING1 first address
LEA DI, STRING2       ; DI←String STRING2 First address
MOV CX, 100           ; CX←Cycles
CYCLE: MOV AL, [SI]   ; AL←characters in string 1
```

```
    MOV BL, [DI]          ; BL←characters in string 2
    INC SI                ; SI←(SI) +1
    INC DI                ; DI←(DI) +1
    CMP AL, BL            ; (AL) - (BL)
    LOOPE CYCLE           ; if (CX) ≠ 0, and ZF =1, turn to CYCLE
    JNZ DONE              ; if the contents of the corresponding two units
                            are not, then turn to DONE
    MOV AL, 0             ; if the two strings are identical, then AL←0
    MOV BL, 0             ; if the two strings are identical, then BL←0
    DONE:…
```

The program uses the LOOPE instruction to control the loop, both counting (CX) control and conditional (ZF) control. There are two possibilities for looping:

① String comparison to find the first different characters: the end of the cycle ZF =0, AL and BL register is the first different characters;

② Comparison string did not find the same character: the end of the cycle ZF =1.

For LOOPZ/LOOPE, LOOPNZ/LOOPNE control of the cycle, the general should be in the end of the cycle with the conditional transfer instructions separate these two cases, respectively.

4. Subroutine call and return instructions

Subroutines have a certain degree of independence to achieve a specific function. The main program in each program need to achieve the specific function can be used CALL instruction call the subroutine, switch to the subroutine execution, the implementation of the subroutine to the end, and then return to the RET instruction to call its main program, continue to follow the instructions, such as Figure 2-42. The subroutine call instruction CALL and the return instruction RET instruction do not affect the flag bit.

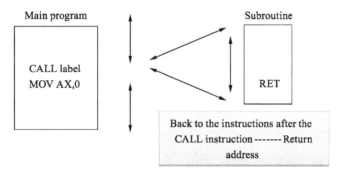

Figure 2-42 Subroutine call and return process diagram

(1) Subroutine call instruction CALL (Call a Procedure)

The subroutine call to the CALL instruction is similar to the JMP instruction, and the execution sequence of the program is transferred by changing the contents of the code segment register CS and the instruction pointer register IP. CALL instruction also supports four kinds of call transfer mode: direct call within the segment, intra-segment indirect call, inter-segment direct call, inter-segment indirect call. The difference from the JMP instruction is that when the subroutine call instruction is executed, the breakpoint address (ie, the current IP or CS: IP content) must be pushed onto the stack protection so that the subroutine execution can be done at the end of the execution, return to the main program to continue. The JMP instruction simply makes the program be transferred without having to return, so the breakpoint address is not saved.

① Section of the direct call

```
CALL near_proc        ; SP←(SP)-2,((SP)+1:(SP))←(IP)
                      ; IP←(IP)+disp
```

The operand of the CALL instruction is the subroutine name. The instruction can be compiled to obtain the relative displacement amount between the next instruction and the entry address of the called subroutine. The relative displacement is 16 bits. Instruction is the first operation of the IP into the stack, and then add the relative displacement of the IP disp, so that control transferred to the subroutine to be called.

② Indirect call within the segment

```
CALL  opr ; SP←(SP)-2,((SP)+1:(SP)) ←(IP)
          ; IP←(opr)
```

CALL pushes the contents of the IP register onto the stack, and then transfers the contents of the operand opr to the IP. The operand opr in the instruction is a 16-bit register or memory location whose contents are the entry address of the proximity subroutine.

③ Between the direct call

```
CALL  far_proc    ; CS stack: SP←(SP)-2,((SP)+1:(SP))←(CS)
                  ; IP stack: SP←(SP)-2,((SP)+1:(SP))←(IP)
                  ; Transfer: CS←far_proc segment address, IP←far_
                    proc offset address
```

The direct call between the CALL directives is to push the current CS and IP values sequentially onto the stack, and then send the remote subroutine segment address and offset address to CS and IP, respectively, so that the control is transferred to the called remote subroutine. The operand in the instruction is a remote subroutine name.

④ Inter-segment indirect call

```
CALL opr          ; CS stack in: SP←(SP)-2, ((SP)+1:(SP)) ←(CS)
                  ; IP stack in: SP←(SP)-2,((SP)+1:(SP)) ←(IP)
                  ; Transfer: CS←(opr) the upper 16 bits, IP←(opr)
                    the lower 16 bits
```

The operand opr can only be a 32-bit memory operand. The CALL instruction first pushes the contents of the current CS and IP into the stack, and then puts the upper 16 bits of the opr content into the CS register, and the lower 16 bits are fed into the IP to transfer to the remote subroutine located in the other code segment.

In actual programming, the assembler automatically determines whether it is a segment or a segment call, and can also use the near ptr or far ptr operator to force a call to a near or far call.

(2) Subroutine return instruction RET (Return from Procedure)

The last executable instruction of the subroutine must be a return instruction that returns to the main program breakpoint of the calling subroutine and continues the main program from the breakpoint.

① There are two types of instructions: return without parameters and return with parameters.

```
RET         ; no parameters, returned within the segment. IP stack: IP←
              ((SP)+1:(SP)), SP←(SP)+2
RET imm16   ; has parameters, returned within the segment. IP stack: IP←
              ((SP)+1:(SP)), SP←(SP)+2
            ; Adjust the stack pointer: SP←(SP)+imm16
```

RET returns the return of the subroutine, and the instruction pops the contents of the top of the stack to the IP register.

② Between paragraphs

There are two types of instructions: return without parameters and return with parameters.

```
RET         ; no parameters, returned between paragraphs. IP stack: IP←
              ((SP)+1:(SP)), SP←(SP)+2
            ; CS stack: CS←((SP)+1:(SP)), SP←(SP)+2
RET imm16   ; has parameters, returned within the segment. IP stack: IP←
              ((SP)+1:(SP)), SP←(SP)+2
            ; CS stack: CS←((SP)+1:(SP)), SP←(SP)+2
            ; Adjust the stack pointer: SP←(SP)+imm16
```

RET returns the inter-segment return of the subroutine, which directs the contents of the top of the stack to the IP register and then pops up the contents of a word unit into the CS register.

RET instruction allows a 16-bit literal parameter imm16, in addition to the pop-up address from the stack pop-up address, but also discards the imm16 specified by the contents of a number of bytes, that is, modify the current stack pointer SP value plus imm16. The with the parameter return instruction is mainly used for the main program through the stack to the subroutine to pass the parameters of the case, RET instruction execution, you can call before pressing some of the parameters to delete the stack. Since the stack operation is a word operation, imm16 is always even.

Note that the type of the RET instruction (within or between segments) is implicit and automatically matches the type of subroutine definition. Although the segment return and the segment return have the same assembly mnemonic, the assembler automatically generates a different instruction code.

[**Example 2-33**]   Call and return example.

```
code  SEGMENT
        ...
        CALL  subp
2000:200H→  ...
        ...
        subp  PROC  NEAR
2000:300H  →  ...
        ...
        RET
        subp  ENDP
code  ENDS
```

Subroutine subp is called before and after the stack changes as shown in Figure 2-43.

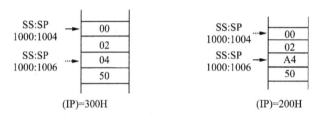

Figure 2-43 Invocation and return stack changes

[**Example 2-34**] Inter-segment call and return example.

```
code1  SEGMENT
        ...
        CALL  far ptr subp
2000:100H→  ...
        ...
        code1  ENDS
        code2  SEGMENT
        ...
        subp  PROC  FAR
3000:200H→  ...
        ...
        RET
        subp  ENDP
        code2  ENDS
```

Sub-program subp is called before and after the stack changes as shown in Figure 2-44.

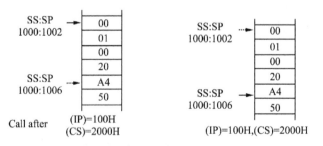

Figure 2-44 Inter-segment call and return stack changes

5. Interrupt and interrupt return command

During the execution of the current program, the CPU suspends the execution of the main program when some exception events or some external request (ie, the interrupt source) occurs, causing the CPU to execute the main program executing the event (ie, the interrupt service Program "). After the execution of the service program, the CPU returns to the main program (that is, the breakpoint) which is suspended. This process is called "interruption". Interrupt is a special way to change the order in which programs are executed. More details on the interruption will be discussed in Chapter 7.

The 8086/8088 CPU supports 256 interrupts, each with an 8-bit number (interrupt type code, also known as interrupt vector number). There are three interrupt instructions: INT, IRET, INTO.

(1) Interrupt command INT (Interrupt)

```
INT n        ; Flag register stack: (SP)←(SP) -2, ((SP) +1, (SP))←
             (FLAGS)
             ; CS Stack: (SP)←(SP) -2, ((SP) +1, (SP))←(CS)
             ; IP stack: (SP)←(SP) -2, ((SP) +1, (SP))←(IP)
             ; Jump: (IP)←(n×4), (CS)←(n×4 +2)
```

The operand n in the INT instruction is an interrupt type code and is an 8-bit literal, which is in the range of 0 to 255. The INT instruction initiates an interrupt service routine with a type code of n. The specific operation is as follows:

① The flag register FLAGS into the stack.
② Clear the flag TF and IF to prohibit single-step interrupt and maskable interrupt.
③ Breakpoint protection: CS into the stack.
④ Breakpoint protection: IP into the stack.
⑤ Realize the jump to the corresponding interrupt service routine (set CS: IP value): Use the interrupt type code n × 4 to calculate the memory unit address where the interrupt service routine entry address is stored, the first word (low address Of the two bytes) into the IP register, the second word (two bytes of high address) into the CS register.

The INT instruction has no effect on the other flags except that the TF and IF are cleared.

From the operation of the interrupt instruction, it can be seen that the entire operation is the same as the indirect call instruction CALL except that the flag register is pushed onto the stack and the interrupt service routine entry address is addressed according to the interrupt type code.

(2) Overflow interrupt instruction INTO (Interrupt of Overflow)

```
INTO; no operand
```

The INTO instruction detects the overflow flag. OF, if OF =1, initiates an interrupt service routine with a type code of 4. Otherwise, no operation is performed, followed by subsequent instructions. When an interrupt occurs, the INTO instruction operates as an "INT 4" instruction.

(3) Interrupt return command (Interrupt Return)

```
IRET       ; restore breakpoint, IP stack: IP←((SP) +1: (SP)), SP←(SP) +2
```

```
; restore breakpoint, IP stack: (SP) +1: (SP)), SP←(SP) +2
; Recovery flag register: FLAGS←((SP) +1: (SP)),SP←(SP) +2
```

The IRET instruction pops the address of the breakpoint pushed onto the stack, returns control to the interrupt call, continues execution of the subsequent instruction, and restores the contents of the flag register. The IRET instruction affects all flag bit states.

Note: The last executable instruction of the interrupt service routine must be IRET.

2.7.5 Processor control instructions

8086/8088 processor control instructions complete the simple control of the CPU, including the flag bit operation, synchronization control and other control and other control functions. In this type of instruction, the rest of the instruction has no effect on the flag except the flag bit instruction.

1. Flag processing instructions

8086/8088 provides seven control flag instructions, you can directly on the flag register FLAGS in the CF, DF and IF bits to operate, to change the status of the flag. See Table 2-15 for details.

Table 2-15 Marking Instructions

Assembly language format	Perform the operation
CLC	Set carry flag, CF = 1
STC	Clear carry flag, CF = 0
CMC	The carry flag is reversed
CLD	Clear direction flag, DF = 0
STD	Set direction flag, DF = 1
CLI	Off interrupt flag, IF = 0, not allowed to interrupt
STI	On interrupt flag, IF = 1, enable interrupts

2. Processor control instructions

Processor control instructions control the processor's working state, do not affect the flag, the following only lists some common instructions, as shown in Table 2-16.

Table 2-16 Processor Control Instructions

Assembly language format	Perform the operation
HLT	Causes the processor to be in a stop state and does not execute instructions
WAIT	So that the processor is in a wait state, TEST line is low, exit waiting
ESC	So that the coprocessor fetches instructions from the system instruction stream
LOCK	Block the bus command, which can be preceded by any instruction
NOP	Empty operation instructions, commonly used in program delay and debugging

2.8 Basic I/O function call

Most of the programs have a "human-machine" interactive process, the keyboard and the display is the "human-machine" interaction of the basic equipment, often requires access to information from the keyboard, and then the computer's processing results through the display output. In assembly language programming, users can directly call the DOS or ROM-BIOS which provides the keyboard and display input and output functions, without having to understand the details of its hardware. This section will be selected in which the commonly used system function calls are introduced.

ROM-BIOS is a set of I/O device drivers that are solidified in ROM. It provides device-level control for the major components of the system and provides character I/O for assembly language programmers operating. Programmers in the use of ROM-BIOS function call, can not care about the hardware I/O interface features, only use the instruction system soft interrupt command (INT n).

DOS system function call is the computer's disk operating system DOS provides a set of routine routines for users. These subroutines can be divided into three main areas: disk read/write and control management, memory management, basic input/output management (such as keyboards, printers, monitors, tape management, etc.), plus time, date and other sub program.

For BIOS and DOS function calls, the general use is the following steps:

① The call parameters into the specified register.

② BIOS or DOS function number loaded AH; if the sub-function number, put it into the AL.

③ DOS or BIOS interrupt (INT) by interrupt type.

④ Check or get the return parameters.

Some function calls do not need entry parameters, then ① can be omitted. After the function call is completed, there are generally export parameters, these export parameters are often placed in the register, through the export parameters, the user can know the success of the call or not.

For example, an ASCII character is displayed on the current cursor position of the screen, and can be called DOS No. 2 function, with the interrupt type number 21H. The block is as follows:

```
MOV DL, '?'        ; Set the entry parameters
MOV AH, 2          ; set function number 2
INT 21H            ; DOS function call
```

The above functions can also be implemented by ROM-BIOS's 0EH function. The interrupt type is 10H. The block is as follows:

```
MOV AL, '?'        ; The characters to be displayed are sent to AL
MOV AH, 0EH        ; function number into AH
INT 10H            ; ROM-BIOS function call
```

2.8.1 Keypad function call (INT 21H)

Table 2-17 lists the main DOS keyboard function calls, including single-character and string input functions.

Table 2-17　DOS keyboard function call (INT 21H)

AH	Features	Call parameters	Return parameter
1	Enter a character from the keyboard and echo it on the screen	NONE	AL = character
6	Read the keyboard character, do not echo	DL = 0FFH	If the characters are desirable, AL = character, ZF = 0; If no characters are available, AL = 0, ZF = 1
7	Enter a character from the keyboard without echoing	NONE	AL = character
8	From the keyboard to enter a character, do not echo, detect Ctrl-Break	NONE	AL = character
A	Enter the character into the buffer	DS:DX = Buffer header	
B	Read the keyboard status	NONE	AL = 0FFH Yes, AL = 00 No typing
C	Clear the keyboard buffer and call a keyboard function	AL = Keypad function number (1, 6, 7, 8 or A)	

1. Single character input

The interactive program often requires the user to respond to a prompt, or by entering a letter or number keys to select the menu items, then use the DOS keyboard call to use single-character input function.

[**Example 2-35**]　Analyze the following block functions.

```
GET_KEY:    MOV AH,1
            INT 21H
            CMP AL,'Y'
            JE YES
            CMP AL,'N'
            JE NO
            JNE GET_KEY
```

If the input is "Y", the program will be transferred to the program labeled YES; and the enter of "N" program will be transferred to the program labeled NO, press the other key program Continue to wait for "Y" or "N" input.

This code can be used in the interactive program, the requirements of the screen display information to answer Y or N occasions.

No. 1 system function call waiting to enter a character from the keyboard, and into the register AL, no entry parameters. Execute the function call 1, the system will scan the keyboard, waiting for a key to press. If the key is pressed, the key value (ASCII value) is read in. If it is Ctrl-Break, if it is, the command is executed. Otherwise, the key value will be sent to AL, and the character will be displayed on the screen.

DOS keyboard call function 1, 6, 7, 8 can enter a character from the keyboard and sent to the AL register, the difference is whether the echo and whether to detect the termination key.

[**Example 2-36**] Compose the program block, check the keyboard input characters for the Enter key, ask to press the Enter key to continue to run.

```
WAIT_HERE: MOV AH, 7      ; function call #7, waiting for keyboard input
INT 21H
CMP AL, 0DH               ; is the Enter key?
JNE WAIT_HERE             ; NO, continue to wait for the next key input
```

When detecting the key input by the keyboard, if the detected keys are letters, numbers or display symbols, etc. that can be written directly in the CMP instruction, enclosed in single quotation marks. But to detect the keys that are carriage return, line breaks and other non-display of control characters, it is necessary to write their instructions in the ASCII code value.

2. String input

The 0AH function of the DOS keyboard call receives a string of characters from the keyboard (ending with a carriage return) and stores it in the user-specified memory input buffer. The first byte of the buffer indicates that the buffer can hold bytes Number (including carriage return), can not be 0. The second byte is initially defined as empty, and after returning from the system function, the actual number of characters entered (not including the carriage return) is entered by the 0AH system function program. From the third byte for the input character storage area, stored from the keyboard to receive the string to the end of the carriage return. If the actual input character is less than the defined number of characters, the buffer will fill the remaining bytes. If the actual input of more than the number of characters defined, the characters will be lost later, and the PC will ring a "beep" sound. When calling, the DS: DX must point to the user-specified buffer.

For example, you need to enter a line of characters, up to no more than 50 (without carriage return), then the input buffer can be defined as follows:

```
DATA SEGMENT
  BUF DB 51           ; buffer length
    DB ?              ; Reserved, fill in the actual number of characters
                        entered
    DB 51 DUP (?)     ; Defines 51 bytes of storage space
DATA ENDS
```

It can also be defined as follows, fully equivalent.

```
DATA SEGMENT
  BUF DB 51,?, 51 DUP (?)
DATA ENDS
```

For the buffer defined above, if the string "ABCDEFG" is input from the keyboard, the contents of each byte of the buffer are: 51, 7, 41H, 42H, 43H, 44H, 45H, 46H, 47H,

0DH,...

[**Example 2-37**] From the keyboard input which is not greater than 65535 decimal number, convert it into binary number, into BINARY.

Analysis: Assuming from the keyboard input "65535 ↙", first of all keyboard input digital character value (ASCII code) into the corresponding value. For example, the decimal numeric character '6' (36H) should be converted to the corresponding value 6. Decimal digital character value and its value difference between 30H, you can use the AND instruction or SUB instruction to achieve the conversion. After the conversion you can get the number: 6, 5,5,3,5.

The numeric size represented by the entered decimal string can be expressed as:
$6 \times 10^4 + 5 \times 10^3 + 5 \times 10^2 + 3 \times 10^1 + 5 = ((((0 \times 10 + 6) \times 10 + 5) \times 10 + 5) \times 10 + 3) \times 10 + 5$

With the binary number of the above operations, you can get 65535 corresponding to the binary number. The above calculation can be calculated by the loop method, once every cycle to complete a "$P = P \times 10 + X_i$" calculation, for the 65535 5 decimal number, you need to repeat the operation 5 times. X_i is the decimal number of each bit value, P initial value is set to 0. The source code is as follows:

```
DATA    SEGMENT
    BUFFER  DB  6, ?, 6 DUP(?)
    C10     DW  10
    BINARY  DW  ?
DATA    ENDS
CODE    SEGMENT
    ASSUME  DS: DATA, CS: CODE
START: MOV AX, DATA
    MOV DS, AX
    LEA DX, BUFFER          ; Load the input buffer first address
    MOV AH, 0AH             ; line input function code
    INT 21H                 ; Enter a number from the keyboard to end with the
                            ;   Enter key
    MOV AX, 0               ; Accumulator is cleared
    MOV CL, BUFFER+1        ; number of cycles
    MOV CH, 0
    LEA BX, BUFFER+2        ; Load the first address of the storage area
ONE: MUL C10                ; P = P × 10
    MOV DL, [BX]            ; remove a character
    AND DL, 0FH             ; converted to binary number
    ADD AL, DL              ; accumulate, P = P × 10 + Xi
    ADC AH, 0
    INC BX                  ; modify the pointer
    LOOP ONE                ; Counting and looping
    MOV BINARY, AX          ; Save the result
    MOV AX, 4C00H
    INT 21H
CODE ENDS
```

END START

2.8.2 Display function call (INT 21H)

Table 2-18 lists the main Dos display function calls.

Table 2-18 DOS display function call (INT 21H)

AH	Function	Call parameter
2	Show a character (check Ctrl − Break)	DL = character, the cursor follows the character movement
6	Show a character (do not check Ctrl − Break)	DL = character, the cursor follows the character movement
9	Display the string	DS:DX = string address, the string must end with $, and the cursor follows the string move

1. Single character output

DOS display function call can be specified in the DL can be given in the DL at the current cursor position, and the cursor moves to the next character position, the difference is whether to detect the termination key.

[**Example 2-38**] The string "Hello, Assembly!" is displayed on the screen.

```
DSEG SEGMENT
    STRING DB 'Hello, Assembly!', '$'
DSEG ENDS
CSEG SEGMENT
    ASSUME CS:CSEG, DS:DSEG
START: MOV AX, DSEG
    MOV DS, AX
    LEA BX, STRING
    MOV CX, 15
NEXT: MOV DL, [BX]        ; set the entry parameters
    MOV AH, 2             ; set function number 2
    INT 21H               ; function call
    INC BX
    LOOP NEXT
    MOV AX, 4C00H
    INT 21H
CSEG ENDS
END START
```

Run the above program, the output string "Hello, Assembly!" May be mixed with other text on the same line output. In order to make the output string occupies a single line of display, you can add some control characters, that is, carriage return (0DH), line breaks (0AH). Therefore, the above procedures in some of the code to do the following changes, the output of the program will be able to exclusive line.

```
DSEG SEGMENT
    STRING DB 0DH, 0AH, 'Hello, Assembly!', 0DH, 0AH, '$'
DSEG ENDS
```

```
...
MOV CX, 19
...
```

Special attention, although the control characters do not show the content, but they are still characters. So the above program to set the number of characters should be set to 19. This function has the same function as Example 2-1, but only calls different system functions to display.

2. String output

DOS display function call function #9 to achieve the output of the string display. When calling, the DS:DX must point to a string in memory with ' $ ' as the end flag. Each character in the string (not including the end flag ' $ ') is output for printing. For example, in Section 2.3.1, Example 2 shows the string "Hello, Assembly!" On the screen, which is called by the 9th DOS display function.

[**Example 2-39**] Outputs the value of a single byte unsigned number in decimal format.

Analysis: Suppose a single byte unsigned X value is 165 (10100101B). To output the display in decimal, the numeric characters '1', '6', '5' should be output in turn. Obviously, as long as we separate the decimal number of the various numerical bits, and then converted to ASCII code can be. So how do you get the decimal number of each bit? The use of "in addition to 10 take the" method can be separated in turn. Need 3 steps:

① Separate the number of digits: first execute X/10, you can get business 16 (0001 0000B), the remainder 5 (0000 0101B), the remainder is separated from the number of single digits 5.

② To separate the tens digit value, that is, the first step of the business of 16 single digits, the same "in addition to 10 to take over", that is, 16/10, get business 1 (0000 0001B), the remaining 6 (0000 0110B) The tens digit of the original decimal number.

③ To separate the number of hundred digits, that is, the second step of a single digit 1, the same "in addition to 10 to take more than", that is, 1/10, get business 0, the remaining 1 (0000 0001B), the remainder is the original decimal The number of hundred digits.

Obviously, the above three steps are the same, which can be achieved by using the loop program, each operation to get the remainder of the decimal number of decimal places 5, 10 digits and 100 digits 1, but the order of the output display should be '1', '6', '5'. The use of the stack of "advanced out of the" operating characteristics, you can solve this problem in reverse order, that is, each step "except 10 to take over" the remainder into the stack, and then display the stack from the pop-up. The last stack of hundreds of bits 1 first out of the stack display.

```
DATA SEGMENT
X DB 165
C10 DB 10
MSG DB 'The Data is:', 0DH, 0AH, '$'
DATA ENDS
CODE SEGMENT
ASSUME DS: DATA, CS: CODE
```

```
        START:
        MOV AX, DATA
        MOV DS, AX
        MOV CX, 3           ; number of cycles, byte unsigned number is less than
                                255, that is, up to 3 decimal digits
        MOV AL, X
        ONE: MOV AH, 0      ; the upper 8 bits are cleared
        DIV C10             ; perform 16b
        8b division
        PUSH AX; Push the remainder (in AH) onto the stack
        LOOP ONE
                            ;Decimal number of bits, ten, hundreds of values have
                                been pushed into the stack.
        MOV DX, OFFSET MSG  ; 9 function entry parameter setting
        MOV AH, 9           ; function number setting
          INT 21H           ; Outputs a string ending in ' $' to the display,
                                outputting a prompt message.
        MOV CX, 3           ; reload CX
        TWO: POP DX         ; pops the remainder from the stack (in DH)
        XCHG DH, DL         ; exchange the remainder to DL
        OR DL, 30H          ; converted to digital ASCII code
        MOV AH, 2
        INT 21H             ; Outputs a character to the display
        LOOP TWO
        MOV AX, 4C00H
        INT 21H
        CODE ENDS
        END START
```

Exercise 2

2.1 8086/8088 CPU which is divided into two major functional components? What is their main function? 8086/8088 What are the registers in the CPU? What are the uses?

2.2 8086/8088 CPU flags in the register which two types of signs? Briefly describe the meaning of each flag.

2.3 What is the difference between the directive "EQU" and " = "?

2.4 Paint Description The following statement allocates the storage space and the initialized value.

(1) FF1 DB '0100', 2 +5,?, 'ABC'

(2) FF2 DW 2 DUP (?), 'A', 'BC', 1000H, 25H

(3) FF3 DB 2 DUP (1,2 DUP (2,3), 4)

2.5 Indicates the following command error.

A1 DB?

A2 DB 10

K1 EQU 1024
(1) MOV K1, AX
(2) MOV A1, AX
(3) CMP A1, A2
(4) K1 EQU 2048
(5) MOV AX, BH
(6) MOV [BP], [DI]
(7) XCHG CS, AX
(8) POP CS

2.6 Assume that the data segment is defined as follows:

```
DATA Segment
XX DB -50, 71, 5, 65, 0
YY DB 200 DUP ('ABCD')
ZZ DW 100 DUP (?)
WW DW 25H, 1052H, 370H, 851H
DATA ENDS
```

(1) Use an instruction to feed the YY offset address into BX.
(2) With a directive to give the data segment occupies all bytes of length.
(3) The number of bytes allocated by the variable ZZ is given by a directive.
(4) Write a program with all the data in the WW array into the YY buffer.
(5) The second data in the array XX and the fifth data exchange.

2.7 What is Addressing? 8086/8088 instruction system which addressing mode?

2.8 Will be the first address of BLOCK byte array of the first 100 number into the AX, try to write the relevant instruction sequence, require the use of the following three addressing modes:
(1) Indirect addressing with BX register.
(2) relative addressing of the BX register.
(3) Addressing with base address of BX, SI register.

2.9 A known, (BX) = 1200H, (BP) = 2400H, (SI) = 0100H, (DI) = 0200H, (SS) = 1000H, (DS) = 2000H, (ES) = 3000H, variable VAR1. Corresponding address is 2000H, indicate the address of the memory in the following instructions and the physical address:
(1) MOV AL, [020H]
(2) MOV AL, [BP +010H]
(3) MOV [BX +SI-20H], AX
(4) MOV BL, ES: [BX +10H]
(5) MOV VAR1 [BX +DI], AL

2.10 Set block for the word unit 1000H: 001FH symbolic address (variable), the contents of the unit is 01A1H, ask the following two contain. Block instructions are different? What is the content of the BX after the instruction is executed?
(1) MOV BX, Block
(2) LEA BX, Block

CHAPTER 2 ASSEMBLY LANGUAGE BASICS

2.11 What is the stack? Using the meaning of the stack.

2.12 If the 0~15 square values are stored in the data segment from the corresponding unit of the byte variable TABLE, try to write the instruction sequence containing the XLAT instruction to find the square of a number in N (0~15). (Set the value of N stored in the CL).

2.13 Prepared by the program in the DX, AX in the double-byte long data complement.

2.14 Write the sequence of instructions that implement the following calculations. (Assuming X, Y, Z, W, R are signed numeric variables).

(1) Z = (W * X)/(R + 6)   (2) Z = ((W-X)/5 * Y) * 2
(3) Z = (X + Y)/R - W   (4) Z = (X/Y + W) * 100 + R

2.15 Installed in the AX, BX, CX, DX are stored in the compressed BCD code that is 4 decimal number, try to write the program to complete the following calculation:

(1) (AX) + (BX) → AX
(2) (DX) - (CX) → DX

2.16 Brief description of the instructions "DAA" and "DAS" on the BCD code after the adjustment of the rules.

2.17 With the block to achieve in the BX, AX double word left to move 5.

2.18 Test analysis of the following procedures to complete what function?

```
MOV CL,4
SHL DX,CL
MOV BL,AH
SHL BL,CL
SHR BL,CL
OR   DL,BL
```

2.19 The known blocks are as follows:

```
MOV AX,1234H
MOV CL,4
ROL AX,CL
DEC AX
MOV CX,4
MUL CX
```

(1) What is the contents of the AX register after each instruction is executed?
(2) What is the value of CF, SF and ZF after each instruction is executed?
(3) What is the value of the AX and DX registers at the end of the program run?

2.20 What is the concept of "transfer" in the program? How does CPU implement the transfer instruction?

2.21 Set(CS) = 600H, (BX) = 0030H, (SI) = 0202H, (20232H) = 00H, (200233H) = 06H, after the implementation of the following two instructions, what is the actual transfer of the target address of the physical address of the number?

(1) JMP BX
(2) JMP Word PTR[BX + SI]

117

2.22 Write the sequence of instructions to achieve the following requirements

(1) The lower 4 bits of the AX register are cleared and the remaining bits are unchanged.

(2) Set the lower 4 bits of the BX register and the remaining bits unchanged.

(3) Test bit 0 and bit 4 in BX, set AL to 1 when the two bits are zero at the same time, otherwise set AL to 0.

2.23 Under what circumstances the implementation of the program is (AH) =0?

```
BEGIN: IN AL, 60H
       TEST AL, 80H
       JZ BRCH1
       XOR AX, AX
       JMP STOP
BRCH1: MOV AH, 0FFH
STOP: ...
```

2.24 What is the difference between the "CALL" instruction and the "JMP" instruction? What is the difference?

2.25 Describe the steps for the 8086/8088 CPU to execute the instruction "CALL DWORD PTR [100H]".

2.26 Set the following procedures before the implementation of the top of the stack pointer SS: SP 1000H: 0220H, try:

```
POP CX
POP BX
POP AX
RET 4
```

(1) Draw the program after the implementation of the stack storage situation diagram.

(2) Gives the current stack top SS and SP values.

2.27 A double word with a long number of symbols placed in X and X +2 (X for the variable), try to write a program for the number of its absolute value.

2.28 Try to write an assembly language program that requires that the lowercase letters entered by the keyboard be displayed in uppercase letters.

2.29 Try to write procedures to achieve the function of Example 2, but do not use the string operation instructions.

2.30 Try to write a program to implement the function of Example 2, but do not use string manipulation instructions.

2.31 What should I do before using the "REPNZ CMPSB" instruction?

2.32 It is known that 100 bytes are stored in the memory area headed by ARRAY, and the relevant blocks are written to complete the transfer of the data to the storage area of the BUFF address. Respectively, with the following different ways to achieve.

(1) With the general data transfer instruction "MOV" to achieve.

(2) Is it implemented by the character transfer instruction "MOVSB".

(3) The instruction "REP MOVSB" is executed with the repeated operation prefix.

(4) With LODSB/STOSB implementation.

2.33 Determine whether the string of length 20 is stored in the memory area of STRING1 and STRING2. If the SIGN unit is set equal, the SIGN unit is set to 0. Respectively, with the following different ways to achieve.

(1) With the comparison of conditional transfer instructions to achieve.

(2) With the repeated operation of the prefix search command to achieve.

2.34 Try other instruction sequences instead of the following instructions:

(1) LOOP NEXT
(2) LDS BX,[100H]
(3) XLAT
(4) LOOPZ NEXT
(5) LAHF
(6) XCHG AX,[BX]
(7) NEG Word PTR[1000H]
(8) ADC AL,[SI+BX]
(9) TEST AL,00001111B
(10) MOVSW(DF=0)
(11) REPZ CMPSB(DF=0)

CHAPTER 3
ASSEMBLY LANGUAGE PROGRAMMING

【Abstract】 This chapter is summarized in order to enable readers to master the basic methods and techniques of assembly language programming, it begins with the simplest sequential programming, from shallow to deep, in order to make it possible for readers to master the basic methods and techniques of assembly language programmed. It introduces branch, cycle programming and subroutine design methods, and also provides corresponding references.

【Learning Goal】
- Master the program structure and its programming methods, including single branch, double branch and other branch programming, counting control cycle, conditional control cycle and other cycle p.
- Get familiar with common programming problems such as array operations, case conversion, parity, character or data count, minimum, maximum, transcoding, and so on.

3.1 Sequential programming

Sequential structure of the program according to the sequence of instructions written in order, no transfer, cycle and other program control class instructions, the order structure is the most basic program structure.

【**Example 3-1**】 Design a program that converts hexadecimal numbers to corresponding seven-segment codes.

The main part of the seven-segment digital tube is a 7-segment LED, which can display hexadecimal numbers (0, ⋯ , 9, A, b, C, d, E, F) by different combinations of seven light.

Each section is controlled by a binary bit of its bright or dark. Therefore, a byte can be used to control the display of seven digital tubes. As shown in Figure 3-1, each section clockwise is called a, b, c, d, e, f, g, and some products are also accompanied by a decimal points h, followed by $D_0 \sim D_7$. Assuming that 0 is the corresponding segment, 1 indicates that the corresponding segment is dark, then the control code corresponding to the number 0 should be 11000000B, the corresponding control code for the

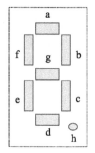

Figure 3-1 seven-segment display digital tube

number 1 is 11111001B, and so on, the corresponding control code for the digital F is 10001110B. This code for controlling the seven-segment digital light is called a seven-segment code.

Obviously the relationship between the hexadecimal number and the seven-segment code is difficult to express as a simple arithmetic expression, so the use of look-up table is to achieve more appropriate code conversion.

The source code is as follows:

```
; Program name: ex301.asm
; Function: hexadecimal number to seven segment code conversion
DSEG SEGMENT
    LEDTB DB 0C0H,0F9H,0A4H,0B0H,99H,92H,82H,0F8H
       DB 80H,90H,88H,83H,0C6H,0C1H,86H,8EH
                              ; Seven paragraph code table
    XDATA DB 9                ; To display the hexadecimal number
    XCODE DB ?                ; Store the seven-segment code to display the
                                corresponding number
DSEG ENDS
CSEG SEGMENT
    ASSUME CS: CSEG, DS:DSEG
START: MOV AX, DSEG
       MOV DS, AX
       MOV BX, OFFSET LEDtb
       MOV AL, XDATA          ; Take hexadecimal numbers
       AND AL, 0FH
       XLAT                   ; Check the table to obtain the corresp
       MOV   XCODE, AL        ; save
       MOV AX, 4C00H
       INT 21H
CSEG ENDS
  END START
```

The key to implement transcoding is the organization of tables. The above procedures in accordance with the size of the hexadecimal number of seven-segment code table, which is easy to find. This code conversion method is simple and quick.

3.2 Branch programming

The branch program is a program that performs different functions according to different conditions or conditions. It has the judgment and transfer function.

There are two elements in the assembly language that implement branches:

(1) Forming conditions: the use of indicators that can affect the status flag, such as arithmetic and logic operations, comparison, testing and other indicators affect the corresponding bit, the status flag is set to correctly reflect the state of the state.

(2) Branch control: the use of conditional transfer instructions, the status of signs such as

testing to determine how to transfer procedures to form a branch.

Branch program structure can have single branch, double branch and multi-branch structure, as shown in Figure 3-2.

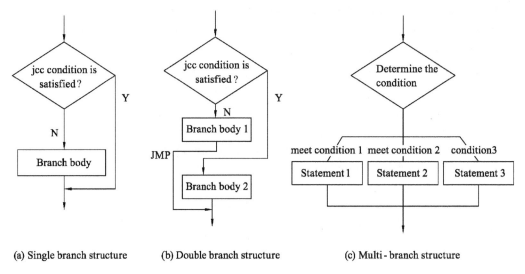

(a) Single branch structure   (b) Double branch structure   (c) Multi-branch structure

Figure 3-2　Branch program structure

3.2.1　Single branch structure

Single branch structure should pay attention to the use of correct conditional transfer instructions. When the conditions are met, then the transfer occurs, skip the branch statement body; if the conditions are not satisfied, then the order of the implementation of branch statements body down. As shown in Figure 3-2 (a).

【Example 3-2】 Calculates the absolute value of the signed number in AL and is stored in the RESULT byte variable.

Analysis: according to the concept of the absolute value in mathematics, the absolute value of a positive number is itself, and the absolute value of a negative number is its inverse. To calculate the inverse of a number, it is necessary to complete the subtraction, that is, minus the number of 0. The 8086 system has a dedicated anti-command NEG. The program flow chart shown in Figure 3-3, is a typical single branch structure, the corresponding block is as follows:

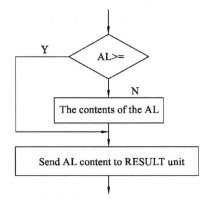

Figure 3-3　Find the absolute value of the program flow chart

```
    CMP AL,0               ; Compare AL with 0
    JGE NONEG              ; Condition to meet (AL≥ 0), transfer
    NEG AL                 ; Conditions are not satisfied, seeking complement
NONEG: MOV RESULT,AL       ; Send the result
```

If you modify the above program in the branch control conditions, the program can be changed to:

```
    CMP AL,0               ; Compare AL with 0
    JL YESNEG              ; Conditional satisfaction (AL<0), transfer,
                             seek complement
    JMP NONEG              ; Condition is not satisfied (AL≥ 0), transfer
                             directly save the structure
YESNEG: NEG AL             ; Condition is satisfied, seeking complement
NONEG: MOV RESULT,AL       ; Send the result
```

Compared to the above two blocks, although they have the same function, the former only conditions meet (AL ≥ 0) will occur when the jump, and the latter whether the conditions are satisfied or not to jump, obviously not as good as the former.

In a single branch program, since only one branch needs to be processed, this part should be followed by the conditional branch instruction, and the part that does not need to be processed is turned away. The condition of the transfer instruction is preferably such that the case does not need to be dealt with set up, jump directly in the past.

3.2.2 Double branch program structure

The double-branch program structure is conditional to satisfy the jump to the execution of the second branch body, and the condition is not satisfied. The first branch statement body is executed sequentially. As shown in Figure 3-2(b).

[**Example 3-3**]   Shows the highest bit of BX.

Analysis: BX the highest bit there are two possible: '1' and '0', to output on the screen display, to be converted to the corresponding ASCII code, that is, 31H and 30H, you can use double branch structure, the corresponding block is as follows:

```
    SHL  BX,1             ; BX the highest bit into the CF logo, you can also use
                            SAL, ROL, RCL instructions
    JC ONE                ; CF=1, That is, the highest bit is 1, transfer
      MOV DL,30H          ; CF=0, That is, the highest bit is 0:DL←30H='0'
      JMP TWO             ; Be sure to skip another branch
ONE: MOV DL,31H           ; DL←31H='1'
TWO: MOV AH,2
    INT 21H               ; display
```

If you modify the above procedures in the branch control conditions, for the double branch structure, as long as the corresponding exchange of two branch statements body can be, the corresponding block is as follows:

```
    SHL BX,1              ; BX highest bit moves into CF flag
    JNC ONE               ; CF=0, that is, the highest bit is 0, transfer
    MOV DL,31H            ; CF=1, that is, the highest bit is 1: DL ← 31H='1'
```

```
        JMP TWO              ; Be sure to skip another branch
   ONE: MOV DL,30H            ; DL←30H = '0'
   TWO: MOV AH,2
        INT 21H              ; display
```

In general, a double branch structure can be designed as a single branch structure. Let us assume that a certain condition is true, write the statement when the condition is satisfied, and then judge the statement and write the statement when another condition is established in a single branch. Note that the appropriate conditional control statement is selected. This example is changed to single branch program structure block which is as follows:

```
        MOV DL,'0'           ; Assume that the most significant bit is '0', DL ← 30H
                               = '0'
        SHL BX,1             ; BX the highest position into the CF logo
        JNC TWO              ; CF = 0, that is, the highest bit is 0, the results
                               have been calculated, direct transfer
        MOV DL,'1'           ; CF = 1, That is, the highest bit is 1, write the branch
                               statement body: DL ← 31H = '1'
   TWO: MOV AH, 2
        INT 21H              ; display
```

In this case to show the numbers 0 and 1, their ASCII value is just a difference of 30H, based on this common, this example can be rewritten into a sequential structure of the program as follows:

```
        MOV DL,0
        SHL BX,1             ; BX the highest position into the CF logo
        ADC DL,30H           ; CF = 0, DL←0 + 30H + 0 = 30H = '0'
                             ; CF = 1, DL←0 + 30H + 1 = 31H = '1'
   TWO: MOV  AH,2
        INT 21H              ; display
```

[**Example 3-4**] Converts a hexadecimal number represented by 4-bit binary to the corresponding ASCII value.

Analysis: hexadecimal number and its corresponding ASCII code have the following relationship:

$$Y = \begin{cases} X + 30H \\ X + 37H \end{cases}$$

Using double-branch structure, the corresponding block is as follows:

```
        MOV     AL,X
                AND     AL,0FH
                CMP     AL,
                JA      ALPH         ; If it is a number between A and F
                ADD     AL,30H       ; If the number is between 0 and 9, the order
                                       is executed
                JMP     DONE
   ALPH:        ADD     AL,37H       ; If it is a number between A and F
   DONE:        MOV     Y,AL
```

CHAPTER 3 ASSEMBLY LANGUAGE PROGRAMMING

Using a single branch structure to achieve, can be rewritten as follows:

```
        MOV   AL,X
        AND   AL,0FH
        OR    AL,30H        ; Let's assume a number between 0 and 9
        CMP   AL,'9'
        JBE   DONE          ; If the number is between 0 and 9, turn DONE,
                              save                           ADD AL,7
                            ;If the number between A~F, plus 7 correction
DONE:   MOV   Y,AL
```

This example can also be used to achieve the transformation of the order structure, that is, using a look-up table to achieve a hexadecimal number converted to ASCII code. Prior to the establishment of a hexadecimal digital characters in the ASCII code table, with the XLAT instruction to perform the escape operation which can complete the required conversion, the specific implementation similar to Example 3-1.

3.2.3 Multi-branch program structure

Multi-branch program structure is a number of conditions corresponding to the respective branch of the body, whose conditions were established on the transfer into the corresponding branch of the body. As shown in Figure 3-2 (c).

[**Example 3-5**] Programming Symbols Functions: Variables X and Y are byte variables in the data segment.

$$Y = \begin{cases} 1 & (X > 0) \\ 0 & (X = 0) \\ -1 & (X < 0) \end{cases} \quad X \text{ range}: (-128 \sim +127)$$

Analysis: this is a 3-branch structure, with two conditional transfer instructions to achieve. The source program flow chart is shown in Figure 3-4. The block is as follows:

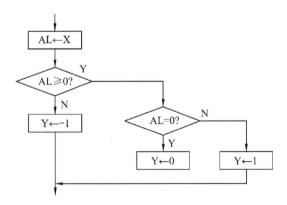

Figure 3-4 Implementation of the symbolic function program flow chart

```
        MOV   AL,X
        CMP   AL,0
        JGE   BIGER
        MOV   AL,0FFH       ; X<0, -1 send to Y   unit
        JMP   OK
```

```
BIGER:    JE    OK           ; X = 0,0 send to Y unit
          MOV   AL,1         ; X > 0,1 send to Y unit
OK:       MOV   Y,AL
```

[**Example 3-6**] Enter the number "1" to "3" from the keyboard, and select the corresponding block execution according to the input.

```
         DATA SEGMENT
             PROMPT  DB    0DH, 0AH, "INPUT A NUMBER (1~3): $"
             MSG1    DB    0DH, 0AH, "FUNCTION 1 EXECUTED . $"
             MSG2    DB    0DH, 0AH, "FUNCTION 2 EXECUTED . $"
             MSG3    DB    0DH, 0AH, "FUNCTION 3 EXECUTED . $"
         DATA    ENDS
         CODE    SEGMENT
                 ASSUME CS: CODE, DS: DATA
         START:  MOV   AX, DATA
                 MOV   DS, AX
         INPUT:  LEA   DX, PROMPT
                 MOV   AH, 9
                 INT   21H          ; Output the message
                 MOV   AH,1
                 INT   21H          ; Enter a number
; **************** ① Start the test conditions ******************
                 CMP   AL, '1'
                 JB    INPUT        ; "0" or non-numeric, re-enter JE F1
                                    ; The number "1", turn F1
                 CMP   AL, '2'
                 JE    F2           ; The number "2", turn F2 CMP AL, '3'
                 JE    F3           ; Number "3", turn F3  JMP
         INPUT                      ; ;Greater than "3", re-enter
; **************** ② Each branch statement sequence ******************
         F1:     LEA   DX, MSG1     ; F1block
                 JMP   OUTPUT       ; At the end of each branch, use the JMP
                                      instruction to jump to the end of all
                                      branches
         F2:     LEA   DX, MSG2     ; F2block
                 JMP   OUTPUT       ; At the end of each branch, use the JMP
                                      instruction to jump to the end of all
                                      branches
         F3:     LEA   DX, MSG3     ; F3block
                 JMP   OUTPUT       ; The last branch ends, JMP can be omitted
         OUTPUT: MOV   AH, 9
                 INT   21H
                 MOV   AX, 4C00H
                 INT   21H
         CODE    ENDS
           END   START
```

Multi-branch structure can be used to achieve a number of conditional transfer instructions, followed by test conditions are met, if the transfer into the corresponding branch of the entrance, if not satisfied, continue to test down until all the test is completed, as shown in Example 3-5 and Example 3-6. This method is simple, intuitive, but slow to run, to check in order to enter the required entry. In order to overcome the weakness of the above method, you can use the address table method to achieve multi-branch, you can directly find the corresponding branch entrance, when the branch has more time, the use of address table method has more obvious advantages. Address table method first in the memory to create an address table, the table in turn store each branch of the entrance address. Example 3-6 uses the address table method, you can do the following changes.

(1) In the data section add a jump table, the table in turn store the entry address of each branch.

```
ADDTBL    DW    F1, F2, F3          ; Note that must be DW type
```

(2) After receiving the numeric input of the keyboard, the block for testing the branch conditions (for example, the block between ① and ② in the source program) can be modified as follows.

```
CMP AL,'1'
JB INPUT            ; Incorrect input, re-enter CMP    AL,'3'
JA INPUT            ; Incorrect input, re-enter SUB    AL,'1'
                    ; Converts the numeric characters "1" to "3" to 0,1,2
SHL AL, 1           ; Convert to 0,2,4
MOV BL, AL
MOV BH, 0           ; go to BX
JMP ADDTBL[BX]      ; Indirect addressing, transferred to the corresponding
                      block
```

3.3 Cyclic programming

In the programming, which often requires a certain program to repeat the implementation of multiple times, this time you can use the loop program structure. In general, a cyclic program structure consists of the following parts:

(1) Loop initialization part: to ensure that the loop can be normal to set the initial value. The initial value of the loop is divided into two categories: one is the initial value of the cyclic working part, such as the accumulator is cleared, the address pointer is set, etc.; the other is the initial value of the control loop end condition. Such as counting the initial value and so on.

(2) Part of the loop: that is, need to repeat the implementation of the program, that is, the body of the loop.

(3) Loop modified part: according to a certain law to modify the operand address and control variables, for the next implementation of the loop body to prepare.

(4) Loop control part: determine the conditions of the loop (the number of loops required or specific loop conditions), decide whether to continue the loop

The loop control can be performed before entering the loop body ("first judgment, post loop"), or after the loop body ("first loop, judgment"). There are two types of looping procedures: the WHILE loop and the DO-WHILE loop, as shown in Figure 3-5.

① WHILE type loop determines the end of the loop conditions, the end of the conditions are not satisfied to enter the loop, the end of the end of the loop is satisfied, so the number of loops may be 0 times.

② DO-WHILE type loop of the implementation of the first loop, and then determine the loop to continue conditions, conditions are satisfied to turn the working part of the loop to continue the loop, the number of loops is at least 1 times.

(5) The end of the loop: mainly used to analyze and store the results of the program.

In particular, the loop initialization and end processing sections are performed only once. In the loop program design must ensure that the loop body and loop control does not turn the loop of initialization part of the statement, otherwise it will cause the death loop or fail to achieve the desired results.

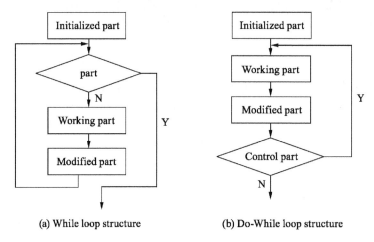

Figure 3-5 Loop structure

How to control the loop is the most important link in the loop program design. There are two most common loop control methods: counting control loop and conditional control loop.

① Count loop: the number of loops is known in advance. The number of loops (called a loop counter) is recorded with a variable (register or memory cell). The count can be either incremental or decrement.

② Conditional loops: the number of loops is not determined in advance, depending on whether a condition is met to determine whether to continue the loop.

3.3.1 Counting cycles

When the loop number is known, usually use count loop. This loop control program with the loop counter value to control loop can also be combined with other common conditions. Control usually uses LOOP/LOOPZ/LOOPNZ/JCXZ instructions to achieve control, use the CX register as the loop counter.

[**Example 3-7**] Byte array ARRAY stored in 10 signed number, find the maximum number of bytes to send variable MAX.

Method for maximum N data: a preset maximum value out of a data are compared with the maximum value, if the data is greater than the maximum value, the maximum value of the data as new. All the data are left after the N data is the maximum value in the default. The maximum value of the initial value can either take one from the N data, and then a number of other N-1 were compared; can also be based on a range of data, within the scope of the decimal number, and then with N. Similarly, it can find several ways to get the minimum value of N data.

In the assembly language, different types of data representation of the scope of the range of the maximum and minimum values are different, the election of the conditional transfer class instructions are also different. The assembly language programmer must know the type of data to be processed, and the default maximum and minimum initial values, as shown in Table 3-1. Where the default maximum number of initial values should be the smallest number within the range, and the default minimum initial value should be the maximum number in the range. For unsigned numbers, they should be transferred with an unsigned conditional transfer class instruction. The signed number should be transferred with a signed number transfer class instruction.

Table 3-1 The maximum and minimum number of the initial value range

Type	8-bit unsigned number	8-bit signed number	16-bit unsigned number	16-bit signed number
Preset the maximum number of initial values	0	−128/80H	0	−32768/8000H
Preset the minimum number of initial values	255/0FFH	127/7FH	65535/0FFFFH	32767/7FFFH
Applicable instructions	JA/JB/JAE/JBE	JG/JL/JGE/JLE	JA/JB/JAE/JBE	JG/JL/JGE/JLE

Analysis: according to the meaning of this case, the first set of the first number of the default number of the current maximum, into the AL, and then the number of AL and the subsequent nine numbers one by one, if the number of AL is small, if the number of ALs is greater than or equal to the number of comparisons, the number of ALs is kept constant, and in the comparison process, the AL is always kept in a large number, In the AL, the final number of AL is into the MAX unit. The data of this question is the number of symbols, should be selected with a symbol number transfer class instructions to achieve conditional transfer. The specific process is shown in Figure 3-6 (Do-While type loop).

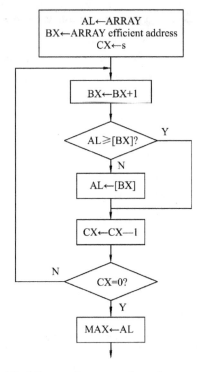

Figure 3-6 Find the maximum number of array program flow

The source code is as follows:

```
DATA  SEGMENT
ARRAY DB -1,59,23,-45,116,107,159,25,218,-14
MAX DB ?
DATA ENDS
CODE SEGMENT
ASSUME CS:CODE,DS:DATA
START: MOV AX,DATA
MOV DS,AX
MOV, AL, ARRAY              ; take the first element of the array as the maximum
                              initial value
MOV BX, OFFSET ARRAY        ; sets the initial value of the address pointer
MOV CX, 9                   ; setting comparisons
LOOP1:INC BX                ; modifies the address pointer and points to the
                              next number to compare
CMP, AL, [BX]               ; comparison
JGE NEXT                    ; the larger the number of AL, the direct end of this
                              comparison, the use of signed number transfer
                              instructions
MOV AL, [BX]                ; the number of AL is small, put the number in AL, so
                              that AL is always the current larger value
NEXT: LOOP LOOP1            ; count cycle control, CX is 0
MOV, MAX, AL                ; compare end, save maximum
MOV AX,4C00H
```

```
INT 21H
CODE ENDS
END START
```

In this case, if the current maximum is preset in the 8-bit signed range, the initial value of the MAX should be set as 80H, starting from the first number of the array, and comparing the 10 to the maximum

3.3.2 Conditional cycles

In some cases, cycle number can not be determined in advance, usually adopts the condition control loop. Some conditions are the implementation cycle and problems, these conditions can be measured through the instruction. If the test results meet the comparison condition, the loop continues; otherwise the end condition of circulation. Circulation is more universal, the essence of count cycle is a kind of special conditions of circulation.

[**Example 3-8**] The sum of the summation of $1 + 2 + 3 + \cdots N$ is completed by programming until the accumulate and exceed 1000. Statistics are accumulated by the number of natural N, accumulate and send SUM. Assume that N and SUM are defined word variables.
Analysis: The number of cycles in this case is uncertain, so the use of conditional cycle control. Process is shown in Figure 3-7, Do-While-type cycle structure.

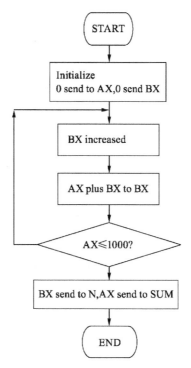

Figure 3-7 N number of cumulative program flow chart

The source code is as follows:

```
FDATA SEGMENT
SUM DW ?
N DW ?
```

```
        DATA ENDS
        CODE SEGMENT
        ASSUME CS:CODE,DS:DATA
        START: MOV AX,DATA
        MOV, DS, AX          ; set DS
        MOV, AX, 0           ; accumulator AX, clear 0
        MOV, BX, 0           ; BX statistics, cumulative number of natural numbers,
                               Qing 0
        LP: INC BX           ; BX plus 1
        ADD, AX, BX          ; find summation
        CMP, AX, 1000        ; compare cumulative and greater than 1000
        JBE LP is less than or equal to 1000, continue to accumulate
                             ; MOV SUM, AX; otherwise, end accumulation, save,
                               accumulate, and
        MOV, N, BX           ; preserve the number of natural numbers accumulated
        MOV AX,4C00H
        INT 21H              ; returns DOS
        CODE ENDS
        END START            ; assembly ends
```

[**Example 3-9**]　　Counts the number of 1 in the BX register and stores the result in the DL register.

```
        The program reads as follows (Method 1):
        MOV AX,BX
        XOR DL,DL
        L:, AND, AX, AX      ; test whether the data in AX is 0, forming a loop
                               judging condition
        JZ EXIT              ; loop control
        SAL AX, 1            ; loop body: implements a count. The highest
                               displacement in AX is entered into CF
        JNC L                ; if CF = 0, turn L
        INC DL               ; if CF = 1, then (DL) + 1 = DL
        JMP L                ; continue to loop at L
        EXIT…
```

The implementation of this function can also be rewritten as a counting loop, and the program segment is as follows (method 2):

```
        MOV AX,BX
        MOV CX,16
        XOR DL,DL
         NEXT:, SAL, AX, 1   ; loop body: implements a count. The highest
                               displacement in AX is entered into CF
        JNC L                ; if CF = 0, turn L at the end of this cycle
        INC DL               ; if CF = 1, then (DL) + 1 = DL
        L: LOOP NEXT         ; loop control: CX minus 1, to determine
                               whether CX is 0, not 0 loops

        EXIT…
```

Analysis: a method using the "cycle control" type WHILE. If (BX) = 0, no circulation, statistical process by shift, shift up to 16 times, there may be only a few times can make shift (BX) = 0, the end of cycle, such as the BX number is 3780H, then as long as the implementation of 9 times. The number of cycles at least 0 times, up to 16 times. The two method is the "count" type DO-WHILE control loop, fixed cycle number is 16 times. For this example clearly method ("WHILE cycle") is more effective.

3.3.3 Multiple cycles

If a loop body contains a cycle, this cycle is called "multi cycle", each layer can be cycle count cycle, can also be a condition of multi cycle cycle. The same design method and single cycle design methods, but should pay special attention to the following points.

(1) The initial control condition and program realization of each cycle.

(2) The inner loop can be nested in the outer loop or nested in multiple layers, but the layers can not intersect the loops, jump from the inner layer to the outer circle, and do not jump directly from the outer loop to the inner loop.

(3) Prevent the occurrence of dead circulation.

The flow chart is shown in Figure 3-8.

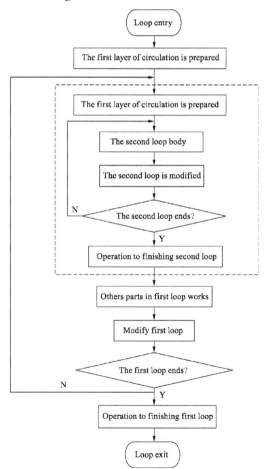

Figure 3-8 Double cycle structure diagram

[**Example 3-10**] A signed byte element array has N signed numbers, which require the N numbers be arranged from small to large.

Analysis: use bubble sort method. From the first data start comparing adjacent numbers, such as the size of the wrong order, the two exchange position. The first time of N-1 times, the largest number has reached the end of the array, only second times N-2 times is enough, in turn, a total of more than N-1 times complete the sort, obviously need double loop. The following source is program design:

```
    DATA SEGMENT
    ARRAY1 DB 15H,0A7H,34H,55H,90H,7EH,3CH,25H,56H,0D6H
    N EQU  $ -ARRAY1
    DATA ENDS
    CODE SEGMENT
    ASSUME CS: CODE, DS: DATA
    START: MOV AX, DATA
    MOV DS, AX
    ******************* began sorting ******************* ;
        MOV, CX, N-1          ; set the outer loop counter; sort the N-1 in CX
    =================== the outer loop is starting =============
    LOOP1: PUSH CX             ; save the external loop counter
        MOV BX, 0             ; BX = the displacement of an ordered element in
                                an array, starting from the first element
                                each time
    ;---------- the inner loop begins, CX is the number of cycles within the
     inner circle ---------------------
    LOOP2: MOV AL, ARRAY1[BX]
    CMP, AL, ARRAY1[BX +1]     ; comparison of adjacent elements
    JLE NEXT                   ; do not need to order, transfer NEXT
    XCHG, AL, ARRAY1[BX +1]    ; swap adjacent element positions
    XCHG AL, ARRAY1[BX]
    NEXT: INC BX               ; modify pointer
    LOOP LOOP2                 ; this time is not over, go to LOOP2, continue
    ;---------- The inner loop body ends ----------
    POP CX                     ; restores the outer loop counter
    LOOP LOOP1                 ; "Times" is not full. Go to LOOP1 and continue
    ================= The outer loop body; regarding the end of ================
    MOV AX, 4C00H
    INT 21H
    CODE ENDS
    END START
```

[**Example 3-11**] There are 4 students taking 5 courses, and the average score of each student and the average score of each course are calculated.

Analysis: according to the problem, the 4 student achievement in proper order is stored in a byte array (the first site for the GRADE), the average score of each student and each class average score were sequentially stored in two byte array (the first site were STU and

COURSE), the following source is program design:

```
DATA SEGMENT
GRADE DB 80,95,76,83,92
DB 65,81,78,84,78
DB 90,86,96,100,83
DB 79,69,88,73,56
STU DB 4 DUP(?)
COURSE DB 5 DUP(?)
DATA ENDS
CODE SEGMENT
ASSUME CS:CODE,DS:DATA
START: MOV AX,DATA
MOV DS,AX
*************** The average score for each student ****************
MOV, DI, 4          ; set the outer loop counter, the number of students in DI
LEA, BX, GRADE      ; sets the address pointer, and BX points to the result
                      table GRADE
LEA, SI, STU        ; sets the address pointer, and SI points to the student
                      average table STU
; ================== the outer loop body starts =================
L11:, MOV, AX, 0; accumulator 0
MOV, CX, 5; set the inner loop counter, the number of courses in CX
---------------------------------- The inner loop body starts ---------------
L22: ADD AL,[BX]
ADC, AH, 0          ; accumulate, consider possible carry
INC BX              ; modify the address pointer
LOOP L22            ; inner loop control
---------- the end of the inner loop, and the total score of 5 courses for a
student in AX ----------
MOV DL,5
DIV DL              ; find the average and save it in AL
MOV, [SI], AL       ; save the average
INC SI              ; modify the address pointer
DEC DI              ; modifies the outer loop count value DI
JNZ L11             ; outer loop control
================== the outer loop body ends ====================
******* The following is the average scores for each course *************
LEA, DI, GRADE      ; sets the address pointer, and DI points to the result
                      table GRADE
LEA, SI, COURSE     ; set the address pointer, and SI point to the course
                      average table COURSE
                    ; MOV DI,BX
MOV, CX, 5          ; set the outer loop counter, the number of courses in CX
================== the outer loop body starts ====================
L01: PUSH CX        ; outer loop count value CX into stack protection
MOV, BX, DI         ; sets the inner loop address pointer
```

```
            MOV, AX, 0         ; accumulate 0
            MOV, CX, 4         ; set the internal loop counter, the number of students in
                                 CX
            ------------The inner loop body starts----
    L02: ADD AL,[BX]
            ADC, AH, 0         ; accumulate, consider possible carry
            ADD BX, 5          ; modify the address pointer and point to the next student's
                                 course address
            LOOP L02           ; inner loop control
            --------------- the end of the inner circle, and the total score of 4 students
            in a course in AX -------------------
            MOV DL,4
            DIV DL             ; asking for GPA in AL
            MOV, [SI], AL      ; save the course average
            INC SI             ; modifies the outer loop COURSE table address pointer
            INC DI             ; modifies the outer loop GRADE table address pointer
            POP CX             ; restores the outer loop count value
            LOOP L01           ; outer loop control
    ; ================== the outer loop body ends ==================
            MOV AX,4C00H
            INT 21H
            CODE ENDS
            END START
```

In the multi-cycle program structure, if the counting loop is used, the processing of the multilayer cycle counter can be used as shown in Figure 3-9. Figure 3-9 lists two approaches. Example 3-10 we use the left-hand processing, Example 3-11 there are two double loops, the first double cycle we use the right way to deal with the way, and the second double cycle using the left side of the way. In addition, the CX register can also be divided into CH and CL two 8-bit registers as the internal and external cycle of the counter, the use of conditional transfer instructions (such as JNZ) control cycle.

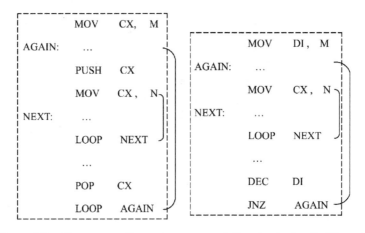

Figure 3-9 The preservation and recovery of Counter in the double cycle

3.4 Subroutine design

The function of relatively independent of the program separately prepared and debug, as a relatively independent module for multiple programs to use to form a subroutine. The subroutine structure can be used to modularize the source program, simplify the source structure, improve code reuse and programming efficiency, especially for complex procedures which are easier to debug and maintain. The main program (caller) needs to call the subroutine (called program) with the CALL instruction. The subroutine needs to return the main program with the RET instruction.

In assembly language, the subroutine uses a pair of procedural directives PROC and ENDP declarations. In general, the subroutine format is basically as follows:

```
; Subprogram list
Subroutine name PROC [NEAR/FAR]
PUSH…protection field (register/memory)
PUSH…the number of decisions based on specific circumstances
… subroutine body
…
POP…to restore the scene, pay attention to the stack order
POP…          ; advanced stack register after the stack
RET           ; return
Subprogram name ENDP
```

Before designing a subroutine, you should first clear the following subroutine list items, which are usually written in annotations before the subroutine code to make it easier for the caller to know how to use the subroutine.

(1) Sub-program name, function and other instructions.

(2) The register or memory unit that is affected in the subroutine.

(3) Subroutine entry parameters, export parameters.

Entry parameter (input parameter): the main program calls the subroutine, provided to the subroutine parameters.

Export parameters (output parameters): subroutine execution ends return to the main program parameters.

(4) The name of the other subroutine is called in the subroutine.

Usually, the main program every time you call a subroutine, you need to do the following three things:

① Prepare the entry parameters for the subroutine.

② Call the subroutine.

③ Handle the return parameters of the subroutine.

[**Example 3-12**]   Find the square root of five unsigned numbers.

Analysis: this example will be an unsigned square root which is designed as a subroutine module ("SQUARE" subroutine), the main program in turn calls the subroutine to obtain the root of five unsigned numbers, the source program is as follows:

```
DATA SEGMENT
X DW 59, 3500, 139, 199, 77        ; an array of squares desired
ROOT DB 5 DUP (?)                  ; Store square root memory area
DATA ENDS
CODE SEGMENT
ASSUME CS: CODE, DS: DATA
START: MOV AX, DATA
MOV DS, AX
LEA BX, X                          ; Initialize the pointer
LEA SI, ROOT
MOV CX, 5                          ; Set the counter initial value
ONE: MOV AX, [BX]                  ; set the entry parameters
CALL SQUARE                        ; call subroutine
MOV [SI], AL                       ; save return parameter (square root)
ADD BX, 2                          ; modify the pointer
INC SI                             ; modify the pointer
LOOP ONE                           ; loop control
MOV AX, 4C00H                      ; End of program execution, return to DOS
INT 21H
; ****************** End of main program ******************
; Name: SQUARE
; Function: Find the square root of the 16-bit unsigned number
; Entry parameters: AX in the desire to set the square root of the unsigned
  number
; Export parameters: AL in the square root
; Impact register: AX (AL)
SQUARE PROC NEAR
PUSH CX; protection site
PUSH BX
; Use the formula: N2 = 1 + 3 + ··· + (2N-1) to find the square root
MOV BX, AX                         ; requires the square root to send the BX
MOV AL, 0                          ; AL stored in the square root, the
                                     initial value of 0
MOV CX, 1                          ; CX Place the first odd 1
NEXT: SUB BX, CX
JB DONE
ADD CX, 2                          ; form the next odd number
INC AL                             ; AL stores the number of odds that have
                                     been subtracted
JMP NEXT
 DONE: POP BX                      ; restore the scene, pay attention to
                                     the order in accordance with the
                                     advanced back to the scene
POP CX
RET                                ; subroutine returns
SQUARE ENDP
CODE ENDS
```

```
        END START                           ; assembly end point
```
Note:

(1) The subroutine is executed by the main program call, so the subroutine should be arranged outside the main program of the code segment. It is usually arranged after the end of the main program execution returns to DOS, the assembly ends before the END directive. Of course, you can also arrange the location in the code before the program starting point.

(2) Subroutine start should protect the contents of the register used and restore it before returning.

(3) Subroutine on the stack of press and pop-up operations is to be used in pairs. Pay attention to keep the stack balance.

(4) Subroutine can share a data segment with the main program, you can also use a different data segment, then pay attention to modify the DS, you can also set the last subroutine data area (using CS addressing).

[**Example 3-13**] Subroutine finally set the data area example.

```
    ; Name: HTOASC
    Function: Converts the hexadecimal number represented by the lower 4 bits
    of AL to the ASCII output display
    ; Entry parameters: AL in the lower 4 bits set and convert the hexadecimal
      number of the display
    ; Export parameters: AL in the conversion of the ASCII code
    ; Influence register: AX
    HTOASC PROC
    PUSH BX
    PUSH DX
    MOV BX, OFFSET ASCII          ; BX points to ASCII table
    AND AL, 0FH                    ; Get a hexadecimal number
    XLAT CS: ASCII                 ; escape code: AL← CS: [BX + AL], data in code
                                     snippet CS
    MOV DL, AL                     ; display
    MOV AH, 2
    INT 21H
    POP DX
    POP BX
    RET
    ;The data area of the subroutine
    ASCII DB '0123456789ABCDEF'   ; hexadecimal number corresponding to the
                                    ASCII table
    HTOASC ENDP
```

(5) Subroutines allow nesting and recursion.

Subroutine itself also called the other subroutine, that is, subroutine nesting. The number of Nested louyers not limited, as long as the stack space is enough, but pay attention to the protection and recovery of the register, to avoiding use of registers between the various subroutines conflict. Subroutine if called subroutine itself, called subroutine recursive call.

[Example 3-14] Subroutine nesting example, in this case the subroutine nested call Example 3-13 subroutine HTOASC.

```
; Name: ALDISP
; Function: Displays the binary number in AL in hexadecimal form
; Entry parameters: AL to display the output of the binary number
; Export parameters: none
; Influence register: none
ALDISP PROC
    PUSH AX          ; protection entry parameters
    PUSH CX
    PUSH AX          ; temporary data
    MOV CL, 4
    SHR AL, CL       ; converts the upper 4 bits of AL
    CALL HTOASC      ; subroutine call (nested)
    POP AX           ; converts the lower 4 bits of AL
    CALL HTOASC      ; subroutine call (nested)
    POP CX
    POP AX
    RET              ; subroutine returns
ALDISP ENDP
```

[Example 3-15] Subroutine Recursive call example.

Programming calculation N!, N! = N(N−1)×(N−2)×2×1 (N >= 0)

Analysis: known recursively defined N! = N×(N−1)!, so, find N! can be designed to input parameters for the recursive subroutine N, each recursive call input parameters decrement 1. If N > 0, the current parameter N is multiplied by the recursive subroutine return value to get the return value of the layer; if the recursive parameter N = 0, the return value is 1. The source code is as follows:

```
    DATA SEGMENT
    N DW 3
    RESULT DW ?               ; Save the result
    DATA ENDS
    CODE SEGMENT
    ASSUME CS: CODE, DS: DATA
START: MOV AX, DATA
    MOV DS, AX
    PUSH N                    ; entry parameter N in stack
    CALL FACT                 ; Subroutine FACT call, find N!
    POP RESULT                ; export parameters pop up to RESULT
    MOV AX, 4C00H             ; return to DOS
    INT 21H
; Name: FACT
; Function: Calculate N!
; Entry parameters: N push into the stack (using stack transfer parameters)
; Export parameters: N! Value at the top of the stack
```

```
; Influence register: none
FACT PROC
PUSH AX
PUSH BP
PUSH DX
MOV BP, SP
MOV AX, [BP+8]              ; takes the port parameters from the stack.

CMP AX, 0                   ; compare whether the entry parameter is 0
JNE FACT1                   ; not 0, then transfer to the call control
INC AX                      ; if 0, set the exit parameter to 0! =1
JMP FACT2
FACT1: DEC AX               ; setting (N-1)! Of the entry parameter values
PUSH AX                     ; entry parameter stack
CALL FACT                   ; recursive call, find (N-1)!
POP AX                      ; export parameters pop up to AX
MUL WORD PTR [BP+8]         ; Find (N)! =N(N-1)!
FACT2: MOV [BP+8],AX        ; set the exit parameters, assuming N! Does not
                            ;   exceed the representation of AX
POP DX
POP BP
POP AX
RET
FACT ENDP
CODE ENDS
END START
```

Analysis: subroutine FACT is a recursive subroutine that takes the stack transfer parameter. Before calling the subroutine to pass a parameter (that is, the factorial N), press onto the stack and call the subroutine FACT. protection. After entering the subroutine, in turn protect the register into the stack, subroutine instruction "PUSH DX" after the implementation of the data stored in the stack as shown in Figure 3-10. According to the figure, in the subroutine read the entry parameters, set the export parameters that are in the [BP+8] stack unit. If seeking 3!, that is, the value of N is 3, the subroutine FACT is called 4 times (the first time by the main program call, the last three times by the subroutine FACT itself call), each call input parameters decrement 1, until decremented to 0 stop recursive call followed by recursive return (backtrack). Each time a call is made, the change in the data in the caller's stack is shown in Figure 3-10. In the figure, the IP value 1 is the offset address value of the instruction "POP RESULT" in the main program, and the IP value 2 is the offset of the instruction "POP AX" (after "CALL FACT") in the subroutine FACT address.

In this case program stack changes are more complex, readers in the analysis of the program must be clear about changes in the stack, or can not grasp the essence of recursive subroutine design. It is advisable for the reader to draw a similar stack diagram when writing such subroutines to avoid stack operation errors.

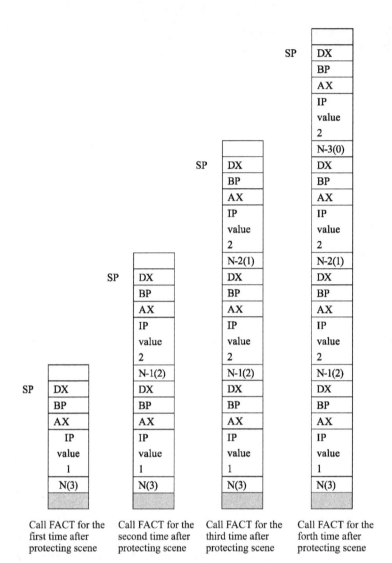

Call FACT for the first time after protecting scene

Call FACT for the second time after protecting scene

Call FACT for the third time after protecting scene

Call FACT for the forth time after protecting scene

CHAPTER 3 ASSEMBLY LANGUAGE PROGRAMMING

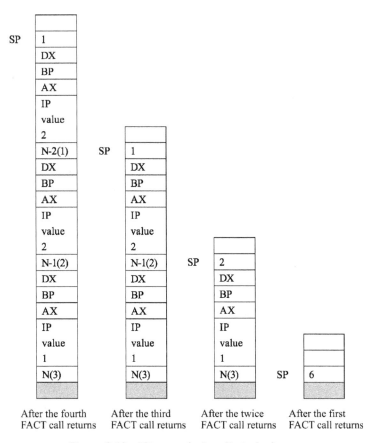

Figure 3-10 N! recursively call stack changes

Exercise 3

3.1   In the block unit headed by Block, a student's test scores are stored. Try to prepare a procedure, the use of student serial number to look up the table to get the student's results, set the student's serial number in the NUMBER unit, look-up table results stored in the RESULT unit.

3.2   The program is programmed to count the number of characters "A" stored in 100 units starting at 52600H and store the result in DX.

3.3   In the current data segment (DS), offset the beginning of the GRADE 80 consecutive units, the deposit of a class of 80 students a test scores. Prepare procedures as required:

① Test preparation program statistics ≥ 90 points; 80 points to 89 points; 70 points to 79 points; 60 minutes to 69 minutes, <60 points for the number of each, and the results on the same data segment, offset address LEVEL Start in the continuous unit.

② Try to write procedures for this class the average score for how much, and placed in the AVER unit of the data segment.

3.4   In the memory area headed by the byte variable ARRAY, a set of signed data is stored, and the trial program adds all positive numbers of the data set and sends them to the

SUM word unit.

3.5 If a class has 50 students, test the assembly language courses, and all the candidates have been stored in the memory unit from the beginning of the unit, try to write a program to find the highest score and the lowest points.

In the first address for the ARRAY address of the memory area, the storage of a set of signed data, try programming.

3.6 Statistics were zero, positive and negative number of the number of statistical results were stored in ZZ, XX, YY.

3.7 A management software can accept ten keyboard commands (A, B, C,···, J, respectively). The program entry addresses for executing these ten commands are PROCA, PROCB, PROCC,···, PROCJ. Write a program to receive commands from the keyboard and go to execute the appropriate program. Requirements are implemented in two ways:

(1) Compare the transfer instruction.

(2) With the address table method.

3.8 Is stored in the register AX, BX, CX in the 16-bit unsigned number, is to write the program, find out the middle of the three values, and put it into the BUFF word unit.

3.9 Write a program that determines whether a binary number x ($2 \leqslant x \leqslant 200$) is a prime number (prime number).

3.10 In the data area headed by BLOCK, the language of the assembly language of a grade (180 person) is stored in the order from small to large. Try to write a program to insert a score of 82 into the appropriate position of the array.

3.11 It is known that array A contains 15 unequal integers, and data B contains 20 unequal integers. It is a program that writes an integer that appears in the array of B and is stored in the array C in.

3.12 The contents of the Flags, AX, BX, CX, DX registers are protected at the beginning of a subroutine, and their contents are restored at the end of the subroutine.

E. g:

```
    PUSHF
    PUSH AX
    PUSH BX
    PUSH CX
    PUSH DX
    ...
    ...
    ; Restore the scene
```

Try to write the sequence of instructions when the scene is restored.

3.13 What is the transfer between the main program and the subroutine? What are the ways to achieve parameter transfer? The application of each method?

3.14 Describes the differences between nested and recursive calls.

3.15 Stores 100 unsigned bytes at address BLOCK. Try to write a program, find the maximum number of the array and the difference between the minimum number, and put it into the RESULT unit, calling the subroutine to complete the maximum number and the minimum number of solutions.

3.16 A grade to participate in the English 4 exam has 250 students, try to write a program to complete 60 −69, 70 −79, 80 −89, 90 −100 four fractional segment of the statistical work, requires the use of subroutine to complete each score section of the statistical work.

3.17 Write a subroutine that performs even parity on the data in AL and returns the checked result back to AL.

3.18 Test program using the subroutine on the subject, the 52600H start 256 units of data plus even parity.

CHAPTER 4
16-BIT MICROPROCESSOR EXTERNAL FEATURES

【Abstract】 This chapter introduces the meaning and characteristics of Intel 16-bit microprocessors 8086 and 8088 two working modes, 16-bit microprocessors 8086, 8088 and 80286 pim name and function. The composition of the 8086,8088 sub system, the bus cycle and the timing.

【Learning Goal】
- Master 8086,8088 maximum and minimum working modes of the composition.
- Master the 16-bit microprocessor pin name and role of various types (transmission signal type, direction and presence or absence of three functions).
- Understand 8086, 8088 maximum and minimum mode of the composition of the microprocessor subsystem to master the function of each component.

4.1 8086 external features

Microprocessor through the pin to complete the external information exchange, different types of microprocessors, the pin settings are not the same. Intel 16-bit microprocessors are mainly 8086,8088 and 80286, the previous chapter first introduced 8086 and 8088 two modes of operation, and then focus on the external features of 8086, the following chapter briefly describes the 8088 and 80286 external features.

4.1.1 8086/8088 working mode

8086/8088 microprocessor accommodates different application environments, there are two modes of operations, namely the minimum mode and the maximum mode.

The so-called minimum mode is the system that has only one 8086/8088 microprocessor. In this case all bus control signals are generated directly from the 8086/8088 pin and the bus control logic in the system is minimized. The minimum mode is suitable for smaller computer applications.

The maximum mode is relative to the minimum mode. In the maximum mode, the system contains at least two microprocessors, of which 8086/8088 is the main processor, other microprocessors known as coprocessors help the main processor to work. In the maximum mode, the bus control signal generated by the 8086/8088 external bus control logic, bus control logic is more complex. The largest mode is used in medium and large-scale microcomputer applications.

There are two types of coprocessors working with 8086/8088, one is the value coprocessor 8087 and the other is the input/output coprocessor 8089. 8087 is a dedicated coprocessor for numerical operations. To achieve a variety of types of numerical operations, such as high-precision integer and floating-point numerical operations, trigonometric functions, logarithmic function calculation. These operations if the software approaches to achieve, will consume a lot of machine time, the introduction of 8087 coprocessor can greatly improve the main processor running speed. The 8089 coprocessor has a set of instruction systems specifically designed for input/output operations that can be directly serviced by the input/output device so that the main processor no longer assumes such work. Adding 8089 coprocessors in the system will significantly improve the efficiency of the main processor, especially in the input/output operation which is more frequently used in the system.

4.1.2 8086 pins

The 8086 microprocessor has 40 pins in dual in-line package, as shown in Figure 4-1.

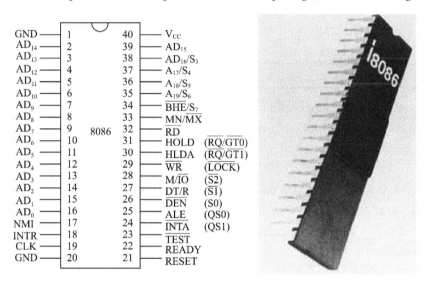

Figure 4-1 8086 pin and package

The following categories describe the 8086 pins.

1. Maximum and minimum mode select pin

The pin is MN/(Minimum/Maximum), the input direction. When this pin is high, the 8086 operates in the minimum mode and operates in the maximum mode when the input is low. In Figure 4-1, 8086 24 ~ 31 pins each have two names, in which the name outside the brackets corresponds to the smallest mode, and the names in parentheses correspond to the maximum mode, that is, the role of these pins in the maximum mode and the minimum mode Is different. The other pins are the same in both modes.

2. Address/data multiplexing pins

Pin multiplexing means that the same pin transmits different types of information at different times. 8086 has 16 address/data multiplexing pins, which are AD15 ~ AD0 (Address/Data), for time-sharing transmission of address information and data information.

When the address information is transmitted, these pins are output and are used to output the address of the memory or peripheral to which 8086 is to be accessed. When transmitting data information, these pins are bidirectional, either to output data or to enter data.

AD15 ~ AD0 has a three-state function, 8086 gives up the control bus, these pins and 8086 within the logic circuit isolation will not affect the other components of the bus control.

3. Address/Status Multiplexed pins

A19/S6 ~ A16/S3 (Address/Status) is the four address/status multiplexed signal output pin, time-out output address and status information, tri-state output. S6 =0 indicates that 8086 is currently connected to the bus; S5 =1 indicates that 8086 can respond to a maskable interrupt request (ie 8086 is currently on, IF =1); the combination of information on S4, S3 is used to indicate the current use of the segment registers, as shown in Table 4-1.

Table 4-1 Relationship between S4, S3 and segment registers

Value of S4, S3	Use the segment register
00	ES
01	SS
10	CS
11	DS

4. Common mode control pin

$\overline{RD}$(Read): read control pin, tri-state output, active low. When active, 8086 reads the memory or peripherals.

$\overline{BHE}/S_7$ (Bus High Enable/Status): high 8-bit data enable/status multiplexed signal pin, time-sharing output $\overline{BHE}$ and status signal S_7. $\overline{BHE}$ =0 indicates that the data on the upper 8-bit data lines D_{15} to D_8 is valid, and $\overline{BHE}$ =1 indicates that the data on D_{15} to D_8 is invalid. The S_7 status signal does not define any practical significance.

Use signal $\overline{BHE}$ and AD_0 signal to determine the current operating type of the 8086 system, as specified in Table 4-2.

Table 4-2 A0 code combinations and corresponding operations

$\overline{BHE}$	A0	Operation	Using the data pin
0	0	Read/write a word AD_{15} ~ AD_0 from the even address unit	AD_{15} ~ AD_0
0	1	Read/write a byte AD_{15} ~ AD_8 from the odd address unit or port	AD_{15} ~ AD_8
1	0	Read/write a byte AD_7 ~ AD_0 from the even address unit or port	AD_7 ~ AD_0
1	1	invalid	
0	1	Read/write a word from the odd address (send the lower 8 bits of data to AD_{15} ~ AD_8 in the first bus cycle and the higher 8 bits to AD_7 ~ AD_0 in the next cycle) AD_{15} ~ AD_0	AD_{15} ~ AD_0
1	0		

READY: ready pin, input, active high. This pin is used to receive a "ready" status signal from the main memory or I/O interface to the CPU. High indicates that the main memory or I/O interface is ready for read and write operations. This signal is the contact signal between the 8086 and the main memory or I/O interface for information transmission.

INTR (Interrupt Request): maskable interrupt request input pin, active high. When this pin is high, it indicates that a maskable interrupt request has been generated externally. If the interrupt enable flag IF = 1, 8086 responds to the request, if IF = 0, it will not respond.

NMI (Non-Maskable Interrupt): non-maskable interrupt request input pin, rising edge is valid. When the pin is changed from low to high, it means that the external non-masked interrupt request is generated and is not restricted by the interrupt enable flag. After the current instruction is executed, the 8086 automatically enters the interrupt service routine.

RESET: reset pin, input, active high. 8086 requires that the reset signal be maintained for at least 4 clock cycles to effect the reset. When the computer cold start or soft start, the reset signal is valid, 8086 reset operation.

$\overline{TEST}$: Test pin, input, active low. The signal on this pin is used in conjunction with the WAIT instruction. The CPU waits after the WAIT instruction is executed. When the pin is input low, the system leaves the wait state and continues execution of the executed command.

5. Minimum mode control pin

$\overline{INTA}$ (Interrupt Acknowledge): maskable interrupt request response pin, tri-state output, active low. The pin output low indicates that the 8086 responds to a maskable interrupt request to inform the interrupt source of the interrupt type number, which is two consecutive negative pulses.

ALE (Address Latch Enable): address latch allows pin, output, active high. When the pin is high, it indicates the current address/data multiplex pin. The address/status multiplexer outputs the address information, and then uses the subsequent falling edge to latch the address information to an external address latch in. The ALE signal can not be floated.

$\overline{DEN}$ (Data Enable): data allows pin, tri-state input, active low. When the pin is output low, it indicates that the 8086 is ready to send or receive data and is typically used as a control signal for the external data bus transceiver.

DT/$\overline{R}$ (Data Transmit/Receive): data transceiver control signal pin, tri-state output. 8086 through this pin to control the direction of data transmission control signal, when the signal is high, that the data from the 8086 output, otherwise 8086 read external data. This signal acts as a second control signal for the external data bus transceiver.

M/$\overline{IO}$ (Memory/Input & Output): memory or I/O port select signal pin, tri-state output. When the pin is high, it indicates that the 8086 reads and writes the memory. When the pin is output low, it indicates that the 8086 reads and writes the I/O port.

$\overline{WR}$ (Write): write control signal pin, tri-state output, active low. And with the realization of the storage unit M/$\overline{IO}$ or I/O port for the write operation control.

HOLD (Hold Request): bus hold request signal pin, tri-state input, active high. The bus request signal is sent to the 8086 for other bus components in the transmission system.

HLDA (Hold Acknowledge): the bus remains responsive to the signal pin, tri-state output, active high. When valid, 8086 recognizes the bus request from other bus components and is ready to let out bus control.

When the 8086 is operating in the minimum mode, the current bus operation type is determined by the sum of the information on the four pins DT/$\overline{R}$, M/$\overline{IO}$, $\overline{RD}$, $\overline{WR}$, as shown in Table 4-3. For example, when the MOV AX, [BX] instruction is executed, the

CPU fetches the memory of a word indicated by DS: BX from the memory and transfers it to the AX register. At this time, the system bus, and the four signal DT/$\overline{\text{R}}$, M/$\overline{\text{IO}}$, $\overline{\text{RD}}$, $\overline{\text{WR}}$ are 0, 1, 0, 1.

Table 4-3 Bus operation types for 8086

DT/$\overline{\text{R}}$	M/$\overline{\text{IO}}$	$\overline{\text{RD}}$	$\overline{\text{WR}}$	OPERATION	BUS SIGNAL
0	0	0	1	Read I/O	$\overline{\text{IORC}}$
0	1	0	1	Read memory	$\overline{\text{MRDC}}$
1	0	1	0	Write I/O	$\overline{\text{IOWC}}$
1	1	1	0	Write memory	$\overline{\text{MWTC}}$

6. Maximum mode control pin

QS_1, QS_0 (Instruction Queue Status): instruction queue status signal pin, output. Their combination gives the status of the instruction queue in the previous T state, facilitating the tracking of the external device's actions on the 8068 internal instruction queue, as shown in Table 4-4.

Table 4-4 QS1, QS0 and state of instruction sequence

QS1	QS0	PERFORMANCE
0	0	No operation
0	1	Fetch code from first byte of instruction sequence
1	0	Sequence is empty
1	1	Fetch codes from following bytes, besides first byte

$\overline{S_2}$, $\overline{S_1}$, $\overline{S_0}$ (Status): bus cycle status signal pin, output. These signals combine to indicate the type of operation performed in the current bus cycle, and the external bus controller uses these signals to generate control signals for memory and I/O ports. The relationship is shown in Table 4-5.

Table 4-5 State code of $\overline{S_0}$ ~ $\overline{S_2}$

$\overline{S_0}$	$\overline{S_1}$	$\overline{S_2}$	PERFORMANCE
1	0	0	Response interruption
1	0	1	Read I/O port
1	1	0	Write I/O port
1	1	1	Stop
0	0	0	Fetch instructions
0	0	1	Read memory
0	1	0	Write memory
0	1	1	No function

$\overline{\text{LOCK}}$ (Lock): bus block signal pin, output, active low. When the pin is output low, other bus components in the system can not occupy the system bus. The $\overline{LOCK}$ signal is generated by the instruction prefix LOCK, and after the execution of an instruction following the LOCK prefix, the signal $\overline{LOCK}$ is canceled. In addition, during the interrupt response bus cycle of the 8086, the signal $\overline{LOCK}$ automatically becomes active low between the CPUs sending two interrupt-response pulses to prevent other bus components from occupying the bus during an interrupt response, resulting in a complete. The interrupt response process is not interrupted.

$\overline{RQ/GT_1}$、$\overline{RQ/GT_0}$ (Request/Grant): bus request/request/enable enable signal pin, bidirectional, active low. When used, the bus request is sent to the external device for the external device. These two pins are available for 8086 other than the two components with the 8086 to negotiate bus usage. $\overline{RQ/GT_0}$ is high priority.

7. Power and clock pins

VCC, GND: power supply, ground pin. 8086 uses a single +5V power supply, connected to VCC, but with two ground pins.

CLK: clock signal input pin, used to input 8086 work when the crystal signal. 8086 maximum clock frequency of 5MHz, the actual application system by the clock chip 8284 to 8086 to provide the clock frequency of 4.77MHz, the duty cycle of about 33% (that is, a cycle of 1/3 of the time for the high level, 2/3 time is low).

4.2　8086 bus operation

Bus operation refers to the 8086 through the system bus and external device information exchange process, such as read and write memory, read and write peripherals and so on. The time required to complete a bus operation is called the bus cycle.

4.2.1　8086 the composition of the bus cycle

8086 all the operations are in the main clock CLK under the control of the beat in an orderly manner, the system clock for a period of time is called the clock cycle. A basic bus cycle of 8086 consists of four T states, which are T_1, T_2, T_3 and T_4, respectively. The duration of each T state is one clock cycle. The bus cycle always starts with T_1 and ends with T_4. If there is no subsequent bus operation after a bus cycle has ended, the bus is in an idle state (T_i). The composition of the bus cycle is shown in Figure 4-2.

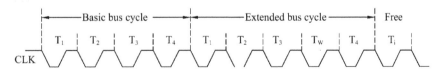

Figure 4-2　8086 bus cycle

8086 in each T state of operation is as follows:

T_1 status: outputs address information and latches to an external address latch.

T_2 state: the withdrawal of the address information, issued a control signal, in order to send data to prepare.

T_3 Status: data is stable if the object being accessed is ready.

T_4 state: read and write data, the end of the bus cycle.

The operating speed of the CPU is usually faster than that of the accessed object. If the bus operation can not be completed by a basic bus cycle, the 8086 can determine whether the bus cycle is extended by the information on the ready pin READY. The specific process is as follows:

8086 always detects the information on the READY pin at the leading edge of the T_3 state. If it is high, it indicates that the accessed object is ready. After the T_3 state, it goes directly to the T_4 state. The bus cycle is not extended and can be transmitted through a basic bus cycle. If the READY pin is low, it indicates that the object being accessed is not ready. After the T_3 status is completed, the wait state T_w is entered and the READY pin continues to be detected at the leading edge of T_w, and it is determined whether or not to continue after the end of T_w. Insert T_w until it detects that the READY pin is high at a T_w edge, then enters T_4 after the end of T_w. A bus cycle containing one or more T_w states is called an "extended bus cycle".

4.2.2 Bus timing of 8086

The bus timing reflects the change in the information on the various types of CPU pins over time in a bus cycle. The following highlights 8086 several typical bus timing.

1. Read bus timing

8086 read memory or read peripherals in many ways is the same, so through a map we introduce the operation of these two bus timing, as shown in Figure 4-3.

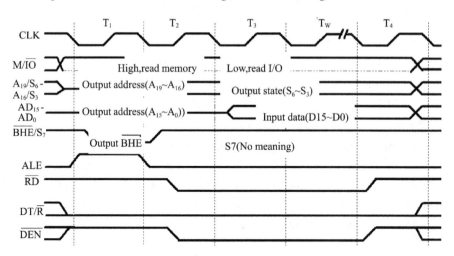

Figure 4-3 8086 read bus timing

The following describes the main functions of each T state.

T_1 status:

① Effective, used to indicate that this read cycle is read memory or read peripherals, it has been maintained until T_4 is valid.

② 20-bit address signal is valid, $A_{19}/S_6 \sim A_{16}/S_3$ send high 4-bit address signal, $AD_{15} \sim AD_0$ send low 16-bit address signal. The address signal is used to indicate the address of the accessed memory cell or the address of the I/O port.

③ Depending on the address characteristics of the data to be accessed (odd or even) and the length (8 or 16), the value of the signal $\overline{BHE}$ is determined and output in the T_1 state.

④ ALE effective. The ALE signal on the trailing edge of T_1 goes low, resulting in a falling edge from high to low. The ALE signal is used as the latch signal for the external address latch and uses the falling edge to latch the address signal sent by the CPU to the address latch.

⑤ When the system is equipped with a data transceiver, T_1 state to become low, used to inform the data transceiver. The bus cycle for the read cycle, in order to receive data to prepare.

T_2 state:

① $A_{19}/S_6 \sim A_{16}/S_3$ send status information $S_6 \sim S_3$.

② $AD_{15} \sim AD_0$ floating, for the back to send data to prepare.

③ $\overline{BHE}/S_7$ starts outputting the status signal S_7, and continues until the end of the bus cycle.

④ $\overline{RD}$ Enable, that to read the memory or peripherals.

⑤ $\overline{DEN}$ Enable, so that the data transceiver can transfer data.

T_3 status:

Data read from memory or peripheral appears on the data bus.

T_4 Status:

In the T_4 and T_3 state of the junction, 8086 collects data, and then revokes the control and status signals, the bus operation is over.

When the bus read operation can not be completed via a basic bus cycle, the 8086 inserts one or several T_w states between T_3 and T_4 according to the READY pin. In the T_w state, each pin is maintained at T_3 unchanged.

2. Write the bus timing

8086 write memory or write peripherals in many ways is the same, so also through a map we introduce the operation of these two bus timing, as shown in Figure 4-4.

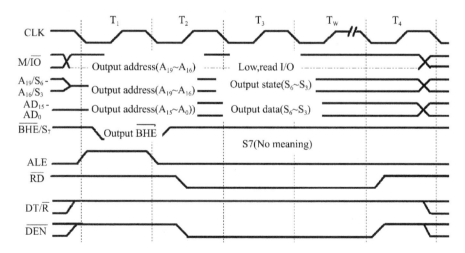

Figure 4-4　8086 write Bus Timing

T_1 status:

Basically the same as the read bus cycle, but at this time output is high but not low.

T_2 state:

There are two main differences between reading the bus cycle:

① $\overline{RD}$ changes $\overline{WR}$, on behalf of the write operation.

② $AD_{15} \sim AD_0$ is not floating, but issued to write to the memory or peripherals in the data.

T_3, T_W, T_4 status is the same as the read cycle.

3. Reset timing

The reset operation of the 8086 is performed by the trigger signal on the RESET pin. When the RESET pin has a high level, the CPU completes the current operation and enters the initialization (reset) process, including the internal register (except CS) 0, the flag register is cleared to 0, the command queue is cleared to 0, and FFFFH is sent to CS. When the RESET from high to low when the trigger a CPU internal reset logic circuit, after 7 T state, CPU is automatically activated. After the restart, the system starts execution from FFFF0H.

When the 8086 is reset, the contents of the internal main registers and instruction queues are shown in Table 4-6.

Table 4-6 State of register after reset

Register	State	Register	State	Register	
FLAG	0000H	IP	0000H	CS	FFFFH
DS	0000H	SS	0000FH	ES	0000H
INSTRUCTION SEQUENCE	EMPTY	IF	0		

The reset timing of the 8086 is shown in Figure 4-5.

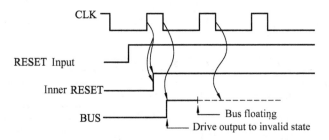

Figure 4-5 8088 Reset Timing

At reset, the control pin output with no tri-state function is inactive and the pins with tri-state functions are all floating.

4.3 8086 microprocessor subsystem

Regardless of whether the 8086 microprocessor is operating in the minimum mode or in the maximum mode, some external chips and processors are required to form a microprocessor subsystem on the outside of the CPU. These peripheral chips provide clock signals for processor operation and provide system data bus, address bus and control bus.

4.3.1 8086 subsystem in minimum mode

Figure 4-6 shows the composition of the 8086 subsystem in the minimum mode, including the 8086 microprocessor, the clock generator 8284A, the address latch 8282, and the data transceiver 8286.

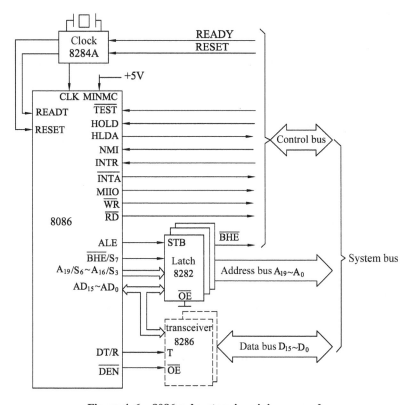

Figure 4-6 8086 subsystem in minimum mode

The clock generator 8284A provides a constant clock signal for the system while synchronizing the READY and RESET signals sent by the external device.

As the 8086 CPU uses the address pin and data pin multiplexing, address pin and state pin multiplexing and other technologies, and in the implementation of the memory read or write or I/O device input and output bus cycle, the memory or I/O device requires that the address information remains active throughout the bus cycle, so that the address latches must be added to form a separate external address bus and data bus when forming the microcomputer system.

In the T_1 state of the bus cycle, the CPU outputs the address information on the multiplexed pin to indicate the address of the memory cell or peripheral port to be accessed, and the CPU sends a high level ALE signal at this time. From the T2 state of the bus cycle, the multiplexed pin is no longer the address information, but the data information ($AD_{15} \sim AD_0$) or the status information ($A_{19}/S_6 \sim A_{16}/S_3$), but because of the address latch pair, address information is latched, so in the second half of the bus cycle, the address and data appear on both the address bus and the data bus. Since the address latch 8282 is an 8-bit latch, the signals to be latched include A_{19}/S_6 to A_{16}/S_3, AD_{15} to AD_0, and $\overline{BHE}/S_7$ are all 21, so three pieces are required. The latch signal STB of the address latch is driven by ALE and the tri-state output enable signal is grounded. The output of the address latch forms the system address buses A19 to A0 and the control bus.

The address/data multiplexing pins AD_{15} to AD_0 are also input as data transceivers 8286, and the outputs form the system data buses D_{15} to D_0. The data transceiver 8286 is an 8-bit bi-directional tri-state buffer that requires two, and their two control signals $DT/\overline{R}$ and $\overline{DEN}$ are driven by the 8086 and pins, respectively. In the $T_2 \sim T_4$ stage of the bus cycle, the data information on the address/data multiplexing pins AD_{15} to AD_0 interacts with the system data bus through the data transceiver 8286. Data transceivers are optional and not essential if the data transceivers are used only if the memory and peripherals connected in the system are larger and the drive capability of the data bus needs to be increased.

In the minimum mode, the system has only one processor, so $M/\overline{IO}$, $\overline{RO}$, $\overline{WR}$, HOLD, HLDA, NMI, $\overline{INTR}$, $\overline{INTA}$ and other control signals are managed by the 8086 processor. $M/\overline{IO} = 0$ and $\overline{RD} = 0$ generate an memory read signal; a combination of $M/\overline{IO} = 1$ and $\overline{WR} = 0$ generate an I/O write. $M/\overline{IO} = 1$ and $M/\overline{RD} = 0$ generate an memory read signal; a combination of $M/\overline{IO} = 0$ and $\overline{WR} = 0$ produces a memory write; signal.

4.3.2 Maximum mode of the 8086 subsystem

Figure 4-7 shows the composition of the 8086 subsystem in maximum mode, including the 8086 microprocessor, the clock generator 8284A, the address latch 8282, the data transceiver 8286, and the bus controller 8288.

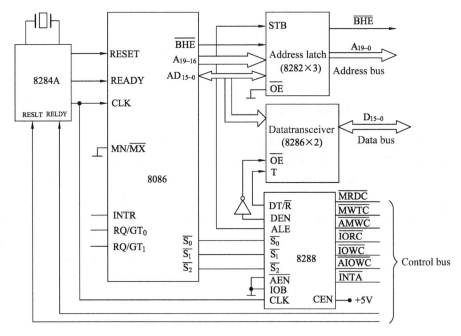

Figure 4-7 Maximum mode of the 8086 subsystem

Comparing Figure 4-6 and Figure 4-7, it can be seen that the main difference between the maximum mode and the minimum mode configuration is in the maximum mode, the bus controller 8288 uses the three status signals $\overline{S_2}$, $\overline{S_1}$, $\overline{S_0}$ issued by the 8086 CPU to decode And then generates a control signal required for the control bus and address latch 8282, data transceiver 8286.

In the maximum mode system, the reason for using the bus controller to transform and combine the control signals is that in a system with a maximum mode, two or more processors are typically included, thus resolving the relationship between the host processor and the coprocessor. The coordination of the work, and the shared control of the system bus, 8288 bus controller plays this role.

The control signal provided by the 8288 mainly has a memory read signal $\overline{\text{MRDC}}$, a memory write signal $\overline{\text{MWTC}}$, an advance memory write signal $\overline{\text{AMWC}}$, I/O read signal $\overline{\text{IORC}}$, I/O write signal $\overline{\text{IOWC}}$, advanced I/O write signal $\overline{\text{AIOWC}}$ and the maskable interrupt request response signal $\overline{\text{INTA}}$.

In the maximum mode system, there are generally interrupt priority management unit 8259A to interrupt the priority of multiple interrupt sources. But if the interrupt source is not much, you can not interrupt the priority management components.

4.4 8088 external features

8088 and 8086 are the same in many ways, so this section will not be described in detail, just 8088 and 8086 different places to introduce.

1. 8088 pin

The 8088 also has 40 address pins in a dual in-line package, as shown in Figure 4-8.

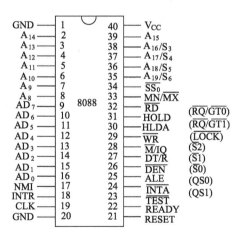

Figure 4-8 8088 pin

8088 and 8086 different places in the performance:

(1) 8088 only has 8 address/data multiplex pins $AD_7 \sim AD_0$, in addition to 8 with address dedicated pins $A_{15} \sim A_8$. Whether it is 8088 or 8086, send the address information pins are 20, can access 1MB memory space. 8088 has eight data pins, and 8086 has 16 data pins, so 8088 can only access 8-bit data each time, and 8086 can access 8-bit data and can access 16-bit data.

(2) 8088 memory or I/O port select signal pin $IO/\overline{M}$, while the corresponding pin $M/\overline{IO}$ of 8086, their role is just the opposite.

(3) 8088 no $\overline{BHE}/S_7$ pin, replaced by $\overline{SS_0}$.

8088 works in the minimum mode, the information on the common control pins DT/R, IO/M and $\overline{SS_0}$ together determines the current operation of the 8088, as shown in Table 4-7.

Table 4-7 8088 bus operation type

DT/R	IO/M	SSO	OPERATION	BUS SIGNAL
0	1	0	Interrupt the response signal	$\overline{INTA}$
0	1	1	Read I/O	$\overline{IORC}$
1	1	0	Write I/O	$\overline{IOWC}$
1	1	1	Stop	—
0	0	0	Fetch instruction	$\overline{IORC}$
0	0	1	Read memory	$\overline{IORC}$
1	0	0	Write memory	$\overline{MWTC}$
1	0	1	No operation	—

2. 8088 bus timing

8088 microprocessor bus cycle configuration and 8086 is the same, the reset timing is the same with the 8086, while the read and write bus cycle and 8086 is basically the same. Figure 4-9 reads the read timing for 8088.

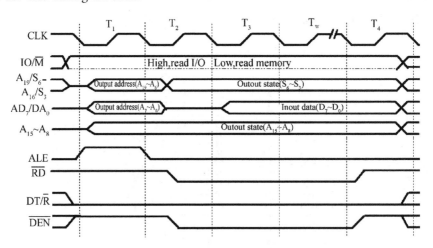

Figure 4-9 8088 read bus timing

8088 read and write bus timing and 8086 read and write bus timing difference is mainly because the two microprocessors are different, reflected in the following three points:

① 8088 pin used to distinguish between access to the object is the memory or peripherals. If the main memory is accessed, the pin is low for the entire bus cycle and is output when the peripheral is accessed and maintained high.

② 8088 has 8 address/data multiplex pin $AD_7 \sim AD_0$ and 8 address dedicated pins $A_{15} \sim A_8$, $AD_7 \sim AD_0$ in the T1 state output address information, in other T state to send data information, and $A_{15} \sim A_8$ in the entire bus Cycle output address information.

③ Since the 8088 does not have the $\overline{BHE}/S_7$ pin, there is no such information in the bus timing diagram.

3. 8088 microprocessor subsystem

8088 minimum mode, the maximum mode of the microprocessor subsystem and 8086 similar, the difference is: address latch need to latch the pin signal including $A_{19}/S_6 \sim A_{16}/S_3$, $A_{15} \sim A_8$ and $AD_7 \sim AD_0$; Data transceiver connected to the 8088 address/data multiplex pin $AD_7 \sim AD_0$, generated by the 8-bit data bus $D_7 \sim D_0$.

4.5 80286 external features

80286 a total of 68 pins, using four flat package QFP (Plastic Quad Flat Package) form, as shown in Figure 4-10.

1. The main pin of the 80286

80286 no longer use the address pin and data pin time-division multiplexing, but has a separate 24 address pins $A_{23} \sim A_0$ and 16 data pins $D_{15} \sim D_0$. Therefore, the 80286 can access the memory space to 16MB; a bus operation can access 8-bit or 16-bit data, the data used to transfer data and A0, which is the same with the 8086.

Figure 4-10 Package of 80286

2. 80286 bus cycle

The signal combination on the 80286 pin, determines the bus cycle type of the 80286, as shown in Table 4-8.

Table 4-8 Bus operation types for 80286

$\overline{CODE/INTA}$	M/IO	$\overline{S_1}$	$\overline{S_0}$	Bus cycle type	Corresponding bus signal
0	0	0	0	Interrupt response	$\overline{INTA}$
0	1	0	1	Read the memory data	$\overline{MRDC}$
0	1	1	0	Write memory data	$\overline{MWTC}$
1	0	0	1	Read I/O	$\overline{IORC}$
1	0	1	0	Write I/O	$\overline{IOWC}$
1	1	0	1	Read memory command (fetch instruction)	$\overline{MRDC}$

* Other combinations are reserved or have no actual meaning.

Exercise 4

4.1 Explain 8086/8088 maximum and minimum working modes definitions and characteristics, how to set the 8086/8088 mode of operation?

4.2 What is the time division of the pin? What are the time-division multiplexing pins of 8086/8088? What are the differences between 8086 and 8088 on the pin?

4.3 Introduces the effect and active level of ALE, HOLD, HLDA, NIMI, INTR and $\overline{INTA}$ pin in 8086 minimum mode, respectively.

4.4 Why set $\overline{BHE}/S_7$ pins in 8086? In the bus cycle of the T_1 state, the pin and AD_0 pin output information which combination of several, respectively, on behalf of what operation?

4.5 8086 and 8088 maximum mode, the minimum mode of the bus cycle type, respectively, by which pin signal to distinguish? 80286 bus cycle type by which pin signal to distinguish? When the 8086 executes the instruction "ADD AX, [BX +10H]", what are the pins, and the output information?

4.6 8086/8088 What are the main components of the microprocessor subsystem in the largest and smallest mode? Briefly describe the functionality of each component. 8086, 8088 microprocessor subsystem, how many pieces of address latch? If necessary, how many data transceiver?

4.7 Brief description of the composition of the 8086, 8088 bus cycle. (DS) =1000H, (AX) =1234H, the main memory access speed is fast enough (do not need to insert TW state), then 8086 in the instruction "MOV [2000H], AX" instruction stage which bus cycle operation? What are the information about the 8086 related pin output in each T cycle of the bus cycle, respectively? If the conversion is 8088 implementation of the above instructions, what is the situation?

4.8 8086/8088 what are the main operations at reset? What is the address of the first instruction executed after resetting?

CHAPTER 5
MICROCOMPUTER INPUT AND OUTPUT TECHNOLOGY

【Abstract】 This chapter is mainly about the classification and function of internal structure, port I/O interface addressing mode and I/O interface technology to read and write, and an instance of the I/O interface, 8-bit and 16-bit I/O, through the I/O interface to exchange data between the host and the several control methods of peripherals, digital input and output.

【Learning Goal】
- Master the function of I/O interface, internal structure and port addressing mode, and understand the classification of I/O interface.
- Master IN/OUT instructions, usage port classification and simple port composition, to understand the I/O interface read and write principle and 8/16 bit I/O organization.
- Understand the process and characteristics of data exchange control between several hosts and peripherals, and master the program design method of program control mode.
- Master the design principle of switch input and output and related program.

5.1 I/O interface overview

The previous chapters mainly explained the host composed of CPU and main memory, and the computer system also needed a large number of various external devices. Computer operator or components can be connected with the computer through them to computer instructions or provide data, in addition, the computer will perform to provide to the operator or external components through them. Hardware devices other than the host are called input/output devices (I/O, Device), or peripherals.

The microprocessor and main memory are connected to each other over a bus, and they are transmitted by binary digits. But there are many kinds of external equipment, different working principles, and great differences in working speed. How can these peripherals be connected with the host computer to form a complete system?

The external device can not be directly connected to the host through the system bus, but add a I/O interface circuit between the host computer and the external device, external devices connected to the system bus through the I/O interface, and the host together constitute a complete hardware system. Different external devices need to be paired with different interface circuits, and all I/O interfaces and peripherals are collectively referred as input and output subsystems. The connection between the host and the peripheral is shown in Figure 5-1.

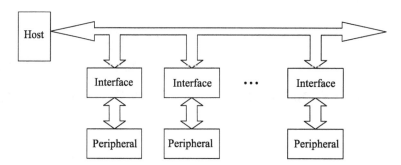

Figure 5-1 Connections between host and peripheral

5.1.1 I/O interface functions

When exchanging information between peripheral and host, the following problems should be solved:

(1) How does the host find the peripheral that exchanges its information from a wide variety of peripherals?

(2) When the work speed difference between peripheral and host is relatively large, how to coordinate?;

(3) How does the host know the status of the peripheral and how to send out control commands?

The solution of these problems is through the I/O interface between the host and peripheral, so the I/O interface plays the role of bridge between the host and the peripheral. CPU does not need to interact directly with peripherals, but rather interact with the I/O interface, shielding the CPU from the variety of peripherals. The capabilities of the I/O interface can be summarized in the following aspects:

(1) Address decoding and device selection. A host can connect to a variety of peripherals, and the host will exchange information with different peripherals at different times. Like access memory, CPU accesses peripherals and sends addresses via the system address bus. After receiving the address information, the I/O interface carries out address decoding, generating device selection signals, and selecting specific devices.

(2) Data buffer function. Because the work speed of host and peripheral device is very different, it is necessary to solve the problem of speed matching between them. In the I/O interface, by setting up one or more data buffer registers for temporary storage of data, the data loss due to inconsistent speed is avoided. When data is transmitted, data is sent into the data buffer and then transmitted to a peripheral or host.

(3) Transmit host control commands and store peripheral status information. When the host and the peripheral exchange the information, needs to understand the peripheral work status information, for example, whether it is free, whether it is prepared, whether it has the breakdown and so on. The host sends corresponding control commands according to the status of the peripherals, such as starting/stopping peripherals, etc.. Set up the corresponding registers in the I/O interface to hold these status information and control commands, which is the

command/status register.

(4) Data format conversion function. In the process of input and output, in order to meet the requirements of the host and peripheral signals, the I/O interface should have the function of information conversion. For example, serial/parallel conversion, serial/serial conversion, digital/analog conversion, analog to digital conversion, etc..

(5) Increasing driving capability and providing working level. One side of the I/O interface is connected to the system bus, but there are many circuits connected on the system bus, and there is a certain transmission distance, so the I/O interface must be able to provide sufficient driving power. The other side of the I/O interface is connected to the peripheral, and the signal level of some peripherals is different from that of the host, so the I/O interface also requires level conversion.

(6) Other functions, such as providing time matching control of host and peripheral, interrupt function and error detection function, etc..

To sum up, the information transferred between the peripheral and host through the I/O interface mainly includes data information, control information and status information.

5.1.2 Composition of interfaces

How can the I/O interface differentiate the data information, status information and control information transmitted between the host and the I/O interface through the data bus of the system? To this end, different registers need to be set up to identify the I/O interface. Figure 5-2 shows the basic components of the I/O interface and the connection between the I/O interface and the host and peripheral devices.

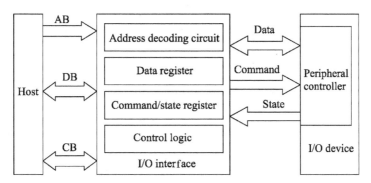

Figure 5-2   Basic structure of I/O interface

An interface usually contains multiple registers, and those registers that can be accessed directly by the CPU are also called ports. Note that interfaces (Interface) and ports (Port) are two different concepts. Several ports are coupled with the corresponding control logic circuit to form an interface. The port is only the register used to store the information in the interface. A register that stores data information is called a data port. Registers that store state information are called state ports, and registers that store control commands are called command ports or control ports.

Each port has a fixed address. As mentioned earlier, CPU access peripherals need to provide address information, which refers to the address of a port in the interface. The address decode circuit in the I/O interface receives the port address sent by the CPU, then generates port control signal by decoding, and selects the corresponding port. The selected ports communicate with the host via the system data bus. Typically, the host can only write to the control port, output control commands, read the status port, and read the status. The data port is divided into data input port and data output port, and the host reads and writes them separately.

5.1.3 Port addressing

CPU must give address information, whether access memory or interface. So, how about the address information sent by CPU, the address of a memory cell in memory, or the address of a port in the interface? This is the problem of addressing by means of I/O port addressing. Common port addressing methods include the following two:

(1) Unified addressing

View the ports in the interface as memory units, and address them uniformly with the main memory unit so that access to the ports is like access to the main memory unit. Using this addressing method, no special I/O instructions are required in the instruction system, and the ports can be operated by the same instructions as memory operations, but the cost is reduced by

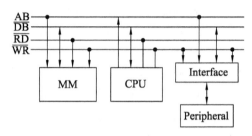

Figure 5-3 Unified addressing

the memory space that can be accessed. Figure 5-3 is a case of unified addressing.

CPU sends address information, read and write control information, to memory and peripherals simultaneously. When the address information for the memory address, the address decoding circuit in memory, the decoding output signal corresponding to the selected storage unit; when the address information for the port address, the address decoding circuit work interface, through the decoding output signal to select the corresponding port.

(2) Independent addressing

In this addressing mode, the I/O port address space and the main memory address space are independent of each other and are individually addressed. Using specialized I/O instructions to access the I/O port, and there is a dedicated signal line to distinguish between current memory operations or I/O port operations. As shown in Figure 5-4. This type of addressing is used in the Intel series.

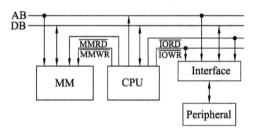

Figure 5-4 I/O ports individually addressed

The address information sent by CPU can also be reached at the same time as memory and interface. If the address is for the memory address, CPU will send a special signal line through the memory read and write control signals, so as to select the appropriate storage unit in the memory; if the address is port address, CPU will send a special signal line through the interface to read and write control signals, and connected to the corresponding port of the selected.

The computer Intel 80x86 series access interface, through the low 16 bits out of port address system address bus, the address bus and the other is invalid, so the I/O address space of 8 port $2^{16} = 64K$ independently addressable components. When accessing the port, you must use the specialized I/O instruction, through the "IN" command to read the port, and use the "OUT" command to write the port. Table 5-1 shows the allocation of port addresses in the 8088/8086 computer system board.

Table 5-1 Assignment of port addresses

I/O address	I/O device port	I/O address	I/O device port
0000 −000F	DMA controller1	200 −207	Game port
0020 −0021	Interrupt controller(main)	0274 −0277	ISA Plug and Play counter
0040 −0043	System clock	278 −27F	Parallel printer port
0060	Control port of keyboard controller	2F8 −2F	Serial communication port1 (COM1)
0061	System speaker	0376	Second IDE hardware controller
0064	Data port of keyboard controller	378 −37F	Parallel printer port1
0070 −0071	System CMOS/real−time clock	3B0 −03BB	VGA display adapter
0081 −0083	DMA controller1	03C0 −03DF	VGA display adapter
0087	DMA controller1	03D0 −03DF	Color display adapter
0089 −008B	DMA controller1	03F2 −03F5	Floppy disk controller
00A0 −00A1	Interrupt controller(Auxiliary)	03F6	First hardware controller
00C0 −00DF	DMA controller2	03F8 −03FF	Serial communication port1 (COM1)
00F0 −00FF	Numerical coprocessor		
0170 −0177	Standard IDE/ESDI hardware controller	No specific port, user can use it.	
001F2 −01FF	Standard IDE/ESDI hardware controller		

5.1.4 Classification of interfaces

From different perspectives, interfaces can be classified differently.

(1) According to the format of data transmission, there are two kinds of parallel interface and serial interface. The parallel interface is a byte (or a word) all the bits at the same time, the serial interface is in the interface between peripherals and transmit the data one by one, and with the host serial interface is still parallel transmission, so the serial interface must be implemented according to the format shift register number.

(2) According to the control mode of host accessing peripherals, there are program query interface, program interrupt interface, DMA interface, and more complicated channel controller. The basic components of these interfaces are described in later chapters.

(3) According to the flexibility of function selection, there are programmable interface and non programmable interface. Programmable interface function and working mode can be changed by the program, to achieve a variety of functions with a programmable interface chip, and interface is not by the program to change its function, can only use the hard wired logic to

achieve different functions.

(4) According to the common component, there are two kinds of general interface and special interface. A common interface is a standard interface for a wide variety of peripherals, and a dedicated interface is designed for certain types of peripherals or for some purpose.

5.1.5 Read and write technology of I/O interface

This section describes how the microprocessor accesses the ports in the I/O interface, including reading and writing. In a computer based on a Intel family of microprocessors, the interface is read/written via the IN/OUT instruction. Let's take 8088 as an example to introduce the problems involved in I/O interface read and write technology.

5.2 I/O interface read and write technology

By executing the IN command, you can read the ports in the interface, and you can write the ports in the interface by executing the OUT command.

5.2.1 I/O instructions

By executing the IN instruction, the port in the interface can be read. By executing the OUT instruction, the port in the interface can be written.

1. Instructions functions and format

The format of the IN instruction is as follows:

```
IN AL/AX, imm8/DX
```

The first operand can only be AL or AX, which means that the contents read to the port are saved to the register inside the microprocessor, AL or AX. The second operand is 8 bit unsigned instant or register DX, used to indicate the port address to read. Such as:

```
IN AL,60H          ; Reads the byte port content of the address 60H and the
                     result is stored in AL
MOV, DX, 200H
IN, AX, DX         ; read the word port content of address 200H, and the
                     result is stored in AX
```

The OUT instruction format is as follows:

```
OUT imm8/DX, AL/AX
```

The first operand is only 8 bit unsigned immediate or register DX, used to indicate the port address to write to. The second operand is AL or AX, indicating the content to be written to the port. Such as:

```
MOV AX,1234h
OUT 40H,AX         ; Change the content of the word port of address 40H to
                     1234H
MOV AL,100
MOV DX,300H
OUT DX,AL          ; Change the content of the byte port of address 300H to
                     100
```

Note: when the port address is greater than 255, the address of the port must be given indirectly by means of DX.

2. Instruction timing

When the microprocessor executes the IN and OUT instructions, an input/output bus cycle is started, and different signal lines in the system bus transmit different information in different T states. The timing of the IN instruction is shown in Figure 5-5(a), and the timing of the OUT instruction is shown in Figure 5-5(b).

At the start of the input/output bus cycle, the address bus address is output on the system address bus, while the $\overline{IO/M}$ signal outputs a high level that represents the access interface, not the memory.

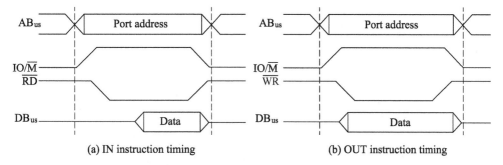

Figure 5-5 Input/output instruction timing

When executing the IN command, the I/O interface sends the contents of the port to the data bus according to the address information, the $IO/\overline{M}$ and the $\overline{RD}$ are selected ports. The microprocessor collects data bus and reads the contents from the ports. When executing the OUT command, the I/O interface stores the data uploaded by the data bus to the port according to the address information, the $IO/\overline{M}$ and $\overline{WR}$ are selected ports.

5.2.2 Port composition

The main function of port is to store information. Ports can be divided into input port and output port according to the direction of transmitting information between port and host. The input port information can only be read out by the host, and the output port information can only be written by the host. Different types of ports have different internal configurations.

1. The composition of the input port

The internal structure of the input port is shown in Figure 5-6. Through the input port, the port can be connected to the input device data and sent to the system data bus, which is read by the host. The three state buffer is the essential device in the input port, which is used to separate the port from the data bus of the system.

The data latch in the input port is used to latch the data information transmitted by the

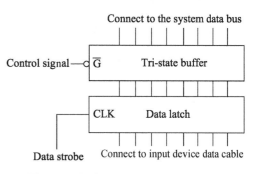

Figure 5-6 Composition of input port

input device. The control signal of the three state buffer is valid only when the host reads on the input port, and its input and output are connected, and the contents of the latch are sent to the data bus of the system. When the wrong input port read operation, control information three state buffer is invalid, input and output is not switched on, the input port and the system data bus isolation, does not affect other parts of the information transmission through the data bus. This is called an input buffer.

When the input device has a data latch function, the data latch in the input port can be omitted, but the three state buffer must be retained. The three-state buffers commonly used in the input port are shown in Figure 5-7.

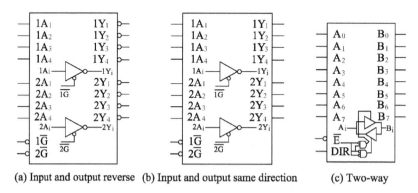

(a) Input and output reverse   (b) Input and output same direction   (c) Two-way

Figure 5-7   common tri – state buffer types

Figure 5-7 (a) for the input and output reverse three state buffers, i. e. the logic level of the input and output of the opposite; Figure 5-7 (b) for the same input and output buffer to the three state logic level input and output the same; Figure 5-7 (c) for two-way three state buffers, information transmission can be performed in two directions in the control of a control the signal.

2. Output ports

The internal structure of the output port is shown in Figure 5-8, and the main component of it is the latch. When the host writes to the output port, the latch signal is valid, and the data uploaded by the system data bus is latched to the output of the latch and is eventually sent to the output device connected to the output port. When a write operation is not performed on the output port, the latch signal is invalid and the output of the latch remains constant. This is called output latch.

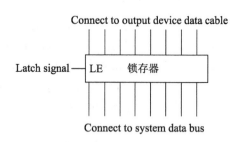

Figure 5-8   Composition of output port

Commonly used latches are mainly shown in Figure 5-9 in several forms. The 74LS273 and 74LS374 are rising edge latches, and the input pin information is latched when the latch signal CLK appears from low to high rising edge. 74LS373 is the falling edge latch, and the input pin information is latched when the latch signal LE appears from high to low falling edge. 74LS373 and 74LS374 also have a three-state output function, only when the control signal is low, the internal latch information can be output through the output pin.

CHAPTER 5   MICROCOMPUTER INPUT AND OUTPUT TECHNOLOGY

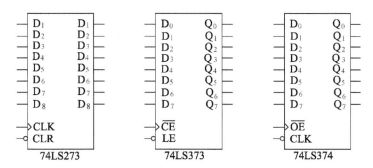

Figure 5-9   Common latches

5.2.3   Address decoding in the interface

The function of the address decoding circuit in the interface is to generate the control signals of each port according to the address information sent by the host and the port read/write control signal. If it is an input port, the output signal of the internal tri-state buffer is generated. If it is an output port, the latch signal of the internal latch is generated. The structure of the address decoding circuit determines the exact address and input/output direction of each port.

Assuming that the control signal of the port is active low, the address of the address decoding circuit given in Figure 5-10 shows that the addresses of the four ports controlled by the address decoding circuit are 330H, 331H, 332H and 333H respectively, and Port 1 and Port 3 can only be read, and Port 2 and Port 4 can only be written.

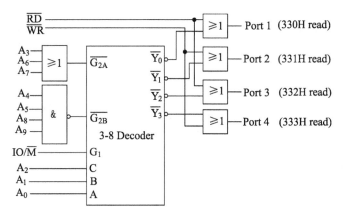

Figure 5-10   Example of address decode circuit

To minimize the number of addresses occupied by each port in the interface, the two ports that are diagonally opposite to the input and output directions are sometimes assigned the same port address, which distinguishes the port by reading the control signal and the write control signal.

5.2.4   Port read and write control

This section introduces the 8-bit port, 16-bit port read and write process through several examples.

1. 8-bit port read and write control

(1) 8-bit port read control

Figure 5-11 in the interface circuit, the bidirectional tri-state buffer 74LS245 as a data input port, with the basic gate circuit address decoding circuit.

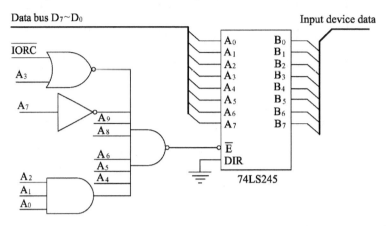

Figure 5-11 8-bit port read operation

When the 74LS245 signal is low, the input device data through the 74LS245 B group of pins and then through the A group of pins on the system data bus. Therefore, to obtain the data of the data device, the address decoding circuit must output a low level, that is, the address information on the system address bus must be 377H, and $\overline{IORC}$ is valid, indicating that the 377H port to read. The program fragment that reads the input device data is as follows:

```
MOV DX,377H
IN AL,DX
```

When the IN instruction is executed, $\overline{IORC}$ is valid, the information output on the address bus is 377H, and the address decoding circuit is enabled so that the input device data appears on the data bus. The microprocessor saves the data on the data bus and deletes the relevant signal, and the reading ends. As the signal is revoked, $\overline{E}$ in the invalid state, 74LS245 input and output pins are not conducting, to achieve system bus isolation, other components can continue to transfer data through the data bus.

(2) 8-bit port write control

In the interface circuit shown in Figure 5-12, the rising edge latch 74LS273 acts as a data output port and consists of a basic gate circuit.

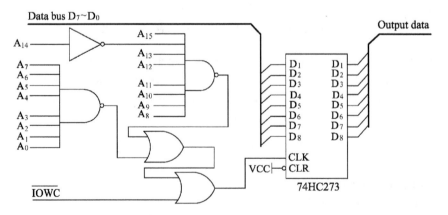

Figure 5-12 8-bit port write operation

When the microprocessor executes the following program segments, the byte data 34H is sent to the output device to which the data output port is connected.

```
MOV AL,34H
MOV DX,0BFFFH
OUT DX,AL
```

When the OUT instruction is executed, the system address bus sends the information BFFFH, $\overline{\text{IOWC}}$ the low level is valid, the information on the system data bus is 34H, and the address decoding circuit outputs the low level. In the latter part of the OUT instruction cycle, $\overline{\text{IOWC}}$ the first becomes high and the address decoding circuit outputs a high level. That is, the output of the address decoding circuit at this time produces a rising edge from low to high, control the 74LS273 to complete the data latch, and start the data on the data bus to the output device. Finally, the address and data are revoked to write data to the port to complete

2. 16-bit port read and write control

Take the 8086 as an example to briefly describe the 16-bit port access operation. Figure 5-13 with two 8-bit buffer 74LS244 as a data input port, the respective eight output pins are connected to 16-bit data bus $D_{15} \sim D_8$ and $D_7 \sim D_0$. Use the basic gate to achieve address decoding.

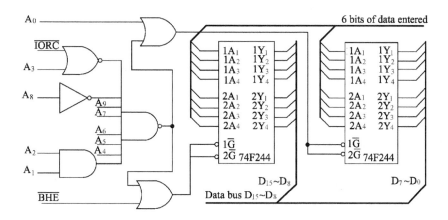

Figure 5-13　16-bit port access

When the CPU executes the following program segments, the 16-bit data of the input device can be read at once.

```
MOV DX,2F6H
IN AX,DX
```

In the implementation of the IN instruction, since the address is even (ie $A_0 = 0$) and 16-bit access (using the register AX), so the output $\overline{\text{BHE}}$ is also active low, two 74LS244 control signals are valid, input device 16-bit data also appears on the data bus $D_{15} \sim D_0$.

5.3 I/O organization

Intel 80x86 computer system, a total of 64K bytes of port, each port has a unique port address. 2 bytes adjacent to the byte port can constitute a 16-bit port, 4 addresses adjacent byte port can constitute a 32-bit port, 8 addresses adjacent byte port can constitute a 64-bit port. Because different microprocessors have different lengths of address bus and data bus, I/O organization is also different.

5.3.1 8-bit I/O organization

8088 computer system uses 8-bit I/O organization, through $A_{15} \sim A_0$ indicate the port address. The organizational structure is shown in Figure 5-14.

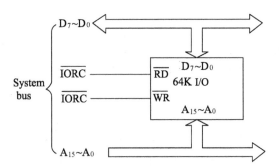

Figure 5-14 8-bit I/O organization form

5.3.2 16-bit I/O organization

8086/80286 computer system uses 16-bit I/O organization form, access port is mainly used in the information $A_{15} \sim A_0$ and $\overline{BHE}$. The entire 64K I/O space is divided into two 32K regions, by $A_0 = 0$ to select the even zone, with $\overline{BHE} = 0$ to select the odd area. The organizational structure is shown in Figure 5-15.

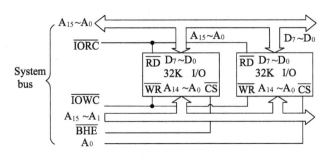

Figure 5-15 16-bit I/O organization form

5.4 Control between the interface and the host information transmission

With the changes in the type of peripherals, the host and peripheral information transmission control mode, from low to high, from simple to complex, from centralized management to the decentralized management of the development process. In this development process, the main four kinds of control methods: program control mode, interrupt mode, direct memory access (DMA) mode and channel mode.

5.4.1 Program control mode

Program control means that the microprocessor through the repeated implementation of the preparation of a good program to complete the access to peripheral information. Program control mode is divided into direct control mode and program query control mode.

1. Direct control mode

Direct control is also known as unconditional transmission. The I/O operation time of some peripherals is known and fixed, and the I/O interface can receive the data to be output from the host at any time, or input data to the host at any time. For example, the CPU output data directly control the indicator light, read the status of the switch directly. Unconditional transmission is only used for simple external devices, the interface is simple, the corresponding interface circuit usually do not need to set the command port, status port and related logic, only need to set the data port and address decoder, as shown in Figure 5-16.

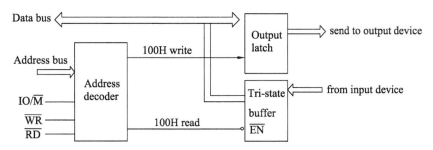

Figure 5-16 Example of I/O interface in direct control mode

With the previous learning base, the internal structure of the address decoding circuit is not shown in Figure 5-16, but a schematic representation is used to represent the data input port (tri-state buffer) connected to the input device in the interface and the port address of the data output port (output latch) connected to the output device is 100H and is divided by the read and write signals. The following program fragment sends the data of the input device to the output device in a direct control mode.

```
MOV DX,100H        ; Specifies the port address
IN AL,DX           ; The data of the input device is read through the data
                     input port
OUT DX,AL          ; The read result is sent to the output device via the
                     data output port
```

2. Program query control mode

Program query control is by the CPU through the program constantly query whether the peripherals are ready to control the host and peripherals exchange information. After starting the peripherals, CPU will continue to query the preparation of peripherals, the termination of the original program execution. This way the CPU and peripherals in a serial working state, CPU efficiency is not high. Only for a small number of peripherals, I/O processing of real-time requirements are not high, and CPU task is a not very busy situation.

(1) the process of querying the program

In the program query mode, the CPU needs to query the status of peripherals. If the peripherals are not ready, CPU must wait and continue to query. If it is already ready, the CPU can make data input or output. So in the interface to set the state port to store the working status of peripherals. The work process of the program query method consists of three basic processes:

① Read the working status of peripheral information.

② Determine whether the peripherals are ready. If not, then return to 1, continue to wait.

③ If it is ready, the cpu starts transmission, modifies pointer and decides whether it is over or not. If the data transmission is not completed, then go to ①, if the thansmission is over, then exit.

The specific program flow of the query is shown in Figure 5-17.

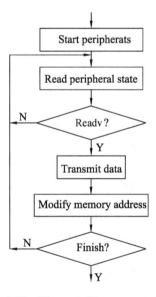

Figure 5-17 The way the program queries

If there are multiple peripherals needed to use the query mode of work, the CPU can be more than one peripheral peripherals query, as shown in Figure 5-18, a total of A, B, C three equipment, CPU rotation one by one, when found outside. If the device is ready, the service is output for the peripheral service. When the peripheral is not ready, or complete the input and output, query the next peripheral, until the last query is completed and then return to the first query, the cycle of bad until all the equipment I/O operations are completed. Throughout the query process, CPU can not do other things. If a peripheral is just in the query after their own in the ready state, then it must wait for the CPU to check the other peripherals again when the query to their own, in order to get CPU services, real-time requirements for relatively high equipment kind of way is not appropriate.

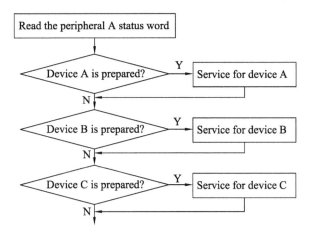

Figure 5-18 Multiple peripherals query the process in turn

(2) Program query mode interface composition

Program query mode interface designs a lot, mainly with the type of system bus, the machine's instruction system and external equipment and other factors. In general, the internal must set the state port, data port and address decoding circuit.

① Query input interface

The input interface circuit of the query mode is shown in Figure 5-19. The latch and the 8-bit tri-state buffer together form the data input port. The D flip-flop and the 1-bit tri-state buffer together form the status port. The 1-bit output signal of the status port is connected to D7 in the system data bus. The data input port and the status port address are 100H, 102H, which can only read.

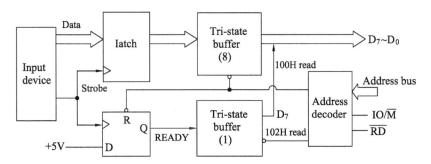

Figure 5-19 Example of Query Input Interface Circuit

After the input device data is ready, a strobe signal is generated. This pulse signal has two functions: on the one hand the data latched into the interface of the data input port, on the other hand the status of the port D trigger set to "1", said the input data is ready. CPU reads the input device data in two steps. The first step is to detect the status bit. Read the status word through the status port and check whether the corresponding READ bit (D_7 bit in the figure) is equal to 1. If READY =1, the peripheral data is already in the data input port. if READY = 0, that is not ready, then continue to detect, then the CPU is waiting for the state. The second step of the implementation of the input instruction from the data input port to read the data, while the D trigger clear "0" for the next data to prepare for the transmission.

The following program fragment reads the data of the peripheral by query.

```
STATUS: MOV DX, 102H       ; Specify status port address
        IN AL,DX           ; Read status port
        TEST AL,80H        ; Test flag D₇
        JZ STATUS          ; D₇ = 0 not ready,continue query
        MOV DX, 100H       ; D₇ = 1 ready,specify data entry port address
        IN AL,DX           ; Retrieve extermal data from data entry port
```

② Query the output interface

The output interface circuit of the query mode is shown in Figure 5-20. The latch constitutes the data output port, the D flip-flop and the 1-bit tri-state buffer form the status port. The 1-bit output signal of the status port is connected to D7 in the system data bus. The data output port address is 100H can only write operation, the state port address is 102H, can only read operation.

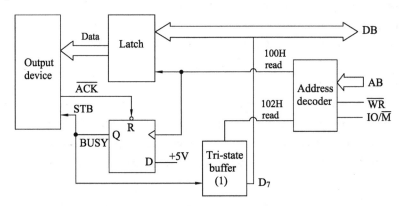

Figure 5-20 Query output interface circuit

When the data is output, the CPU can send the new data to the data output port in the interface only if the peripheral is in the idle/busy state (BUSY =0). Therefore, the CPU to the external output data should first detect the peripheral state. Read the status word through the status port and check whether the corresponding BUSY bit (D7 bit in the figure) is equal to 0. If BUSY =1, it indicates that the peripheral is still busy. The data sent to the data output port Not removed by the peripherals, CPU can only continue to query, wait, until BUSY =0. If BUSY =0, it indicates that the data in the data output port of the interface has been removed by the peripheral. The CPU sends the new data to the data output port of the interface by executing the output instruction, and sets the BUSY trigger to "1". When the output device takes the data sent by the CPU, the interface back to send a response signal, so that the BUSY trigger clear "0" for the next data transfer to prepare.

The following program fragment is used to query the output to the peripheral output variable CHAR in an 8-bit data.

```
STATUS: MOV DX, 102H       ; Specifies the status port address
        IN AL,DX           ; Read status port
        TEST AL,80H        ; Test flag D7
        JNZ STATUS         ; D7 =1,Busy, continue to query
        MOV DX, 100H       ; D7 =0,Not busy, specify the data output port
        MOV AL,CHAR
        OUT DX,AL          ; Write new data to the data output port
```

From the above description it can be seen in the program query mode, that the CPU is completely occupied by the peripherals, and can not do anything else.

5.4.2 Program interrupt mode

After the host starts the peripherals, it is no longer inquiring whether the peripherals are ready, but continue to execute their own programs. When the peripherals are ready to send an interrupt request to the host, the CPU responses to interrupt requests after the implementation of a good preparation of the input and output procedures, through the program input and output instructions to complete the data transmission, and then return to continue to implement their own procedures. Obviously, the use of interrupt mode will greatly improve the efficiency of the CPU.

Program interrupt mode is the most important way in several control modes, it not only allows the host and peripherals to work in parallel, and allows a host to manage multiple peripherals, so that they work at the same time. But the completion of a program interrupt also need a lot of auxiliary operations, when the number of peripherals is too much, too many interrupt requests may also make the CPU too late to respond. Some other high-speed peripherals, because the exchange of information is in batches, if the processing is not timely, may result in loss of information. Therefore, the program interrupt mode is mainly applicable to low, medium speed peripherals. A detailed description of the interruption is given in Chapter 7 of this book.

5.4.3 Direct Memory Access (DMA) mode

Compared with the program query mode, the program interrupt mode greatly improves the efficiency of the CPU, and the CPU has the ability to deal with emergencies. However, when the speed of peripherals to further improve, even close to the speed of execution of an instruction, the CPU utilization will be reduced. The reason is that every time an interrupt is interrupted, the program must be switched once, and an interrupt-related operation consisting of several instructions must be performed. But if excessive interruptions are bound to reduce CPU utilization, and may even cause data loss. So for high-speed I/O devices, you must find a more efficient input and output, this way is the DMA.

DMA mode is Direct Memory Access (Direct Memory Access), also is a direct dependence on hardware in the main memory and I/O devices for data transfer between the I/O data transfer control. The hardware that controls the data transfer process is called a DMA controller (DMAC). DMA mode in the process of data transmission does not require CPU intervention does not require software intervention, usually used for high-speed peripherals in accordance with the continuous address to access the main memory of the occasion.

Compared with the program interrupt mode, DMA mode only need to take up the system bus, do not need to switch programs, thus saving a lot of CPU time, making the CPU efficiency improved.

DMA mode, the realization of data transfer directly by the hardware control, improve the efficiency of data transmission, but also increased the complexity of the DMA interface. DMA mode can only achieve data transfer, a single function, not like the way to interrupt the software to achieve a variety of complex functions.

DMA mode is generally used in high-speed peripherals and main memory between the simple data transmission, and other high-speed data transmission related occasions. For example, for dynamic memory refresh; for disk, tape, CD-ROM and other external memory device interface; for network communication interface; for high-volume, high-speed data acquisition interface.

As the DMA method itself can not handle more complex things, so in some applications often DMA mode and program interrupt mode integrated application, the two complement each other. See Chapter 9 for a detailed description of DMA.

5.4.4　Channel mode

Channel mode is the further development of DMA. Setting up a separate component in the computer system that can replace the CPU management control peripherals, the I/O channel, is a controller that can perform limited channel instructions to achieve higher parallelism between the host and the peripheral. Host in the implementation of I/O operation, only need to start the relevant channel, the channel will perform the channel program to complete the input and output operations. A channel can control multiple different types of peripherals.

Small, microcomputers are mostly used in the first three ways, large, medium-sized machine to use more channels. In recent years, the medium and small and micro-computer also developed this technology, the formation of a variety of I/O processor. And in the commitment to high-end computing large, giant machine, the extensive use of peripheral processors.

5.5　Digital input and output

Switch input and output is not subject to state constraints, it is usually used unconditionally direct control of the input and output methods, which can be completed through a simple interface circuit.

5.5.1　Switch output

Common switch output applications are two: LED light-emitting diode control and implementation of component drive coil. The following LED light-emitting diode as an example to introduce the switch output.

Light-emitting diodes are often used to indicate the state of the instrument and can be driven directly by logic circuits. Common LED digital tube is composed of a number of light-emitting diodes, each piece of information corresponds to a light-emitting diode, as shown in Figure 5-21.

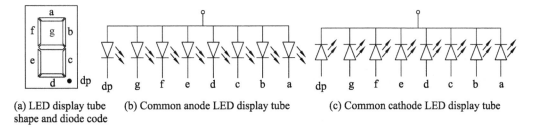

(a) LED display tube shape and diode code   (b) Common anode LED display tube   (c) Common cathode LED display tube

Figure 5-21   LED digital tube

For common anode digital tubes, all segments of the anode are connected together and connected to high (logic 1). When a bit input is 0, the corresponding light-emitting diode is lit, the input is 1 off. Common cathode digital tube is just the opposite. To display specific information on the digital tube, you must enter a specific information from the input code, called the font code, also known as segment code.

To drive the LEDs, simply set a simple interface between the diode and the host that contains only the data output port and address decode circuit, as shown in Figure 5-22. The control signal of the latch is generated by the address decoding circuit.

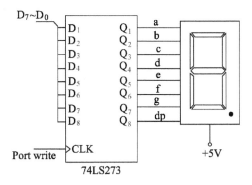

Figure 5-22   switch output interface

If the digital tube is a common anode and the system data lines $D_7 \sim D_0$ are connected to the $D_8 \sim D_1$ pins of the 74LS273 in order from the high to the low order, the font codes of the 0 to 9 ten-word information are shown in Table 5-2.

Table 5-2   0 ~ 9 common anode word code table

Font information	0	1	2	3	4	5	6	7	8	9
Font code	C0H	F9H	A4H	B0H	99H	92H	82H	F8H	80H	90H

If the address decoding circuit in the interface generates a valid port write signal when the address information is 200H, the following program will display the message "3" on the digital tube.

```
        MOV DX,200H
        MOV AL,0B0H
        OUT DX,AL
```

The following is a common anode digital tube, for example, a number of digital tube on the display of information on the method. If you want to use 8 common anode digital display information, a relatively simple method is: 8 digital tube segment code input signal in parallel, by the interface circuit in an output port drive (that is, through the port to send the font code can be received by multiple digital tubes, called the segment port), and each of the other output port control each digital tube of the common anode signal (the port of a bit is 1, the connected Digital display shows the paragraph code information, called the bit code port). Multi-digit digital display interface shown in Figure 5-23.

The general process of multi-digit digital display program is:
(1) Set the bit code be invalid, extinguish all digital tube.
(2) Will be a digital tube code into the segment code port.
(3) Set the bit code, lit a digital tube.
(4) After the appropriate delay, repeat the above process.
Figure 5-23 digital display interface.

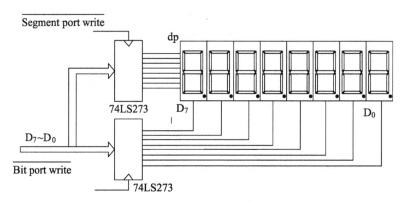

Figure 5-23 Multi-digit LED display interface

If the data bus $D_7 \sim D_0$ from the high to the low bit through the segment code port in turn connected to the digital tube font code input pin dp, g, f, e, d, c, b, a, and through the bit code The port is connected in turn from left to right 8 digital tube common anode pin, then the following program function is in the 8 digital tube from left to right in order to show 1 to 8.

```
        .DATA
        SEGTAB DB 0C0H,0F9H,0A4H,0B0H,99H,92H,82H,F8H,80H,90H
        ; 0 ~ 9 Segment code table
        BUFFER DB 1,2,3,4,5,6,7,8
        SEGPORT DW ?
        BITPORT DW ?
        BITCODE DB ?
        .CODE
START: CALL LED_DISP
        MOV AX, 4C00H
```

CHAPTER 5 MICROCOMPUTER INPUT AND OUTPUT TECHNOLOGY

```
            INT 21H
    LED_DISP PROC FAR
        ...
        MOV AX, @ DATA
        MOV DS, AX
        LEA BX, SEGTAB
        MOV BITCODE, 80H
        MOV SI,0
        MOV CX,8
    ONE: MOV AL,0
        MOV DX,BITPORT
        OUT DX,AL
        MOV AL,BUFFER[SI]
        XLAT
        MOV DX,SEGPORT
        OUT DX,AL
        MOV AL,BITCODE
        MOV DX,BITPORT
        OUT DX,AL
        ROR BITCODE, 1
        INC SI
        ; Appropriate delays should be applied
        LOOP ONE
        ...
        RET
    LED_DISP ENDP
    END START
```

5.5.2 Switch input

Common input switch is mainly in three forms, single-pole single-throw switch, single-pole double-throw switch and button, respectively, as shown in Figure 5-24 below.

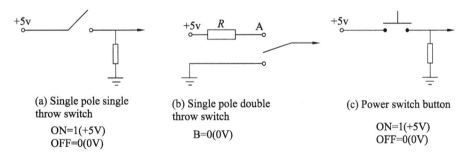

(a) Single pole single throw switch
ON=1(+5V)
OFF=0(0V)

(b) Single pole double throw switch
B=0(0V)

(c) Power switch button
ON=1(+5V)
OFF=0(0V)

Figure 5-24 Three kinds of forms of switch

When the switch position (open or closed), the button state changes (press or release), the switch circuit output level changes, read the switch amount is different.

To read the switch, only in the switch between the circuit and the host set only contains data input port and address decoding circuit can be a simple interface, and the data input port only three-state buffer without the need for latch, as shown in Figure 5-25. The control signal of the tri-state buffer is generated by the address decoding circuit. From the data input port to read the information, each binary bit directly reflects the switch position, when the "1" when the disconnect, "0" when closed.

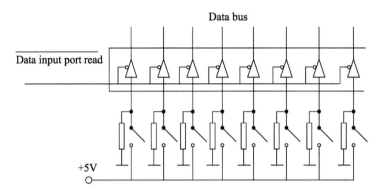

Figure 5-25 Binary input interface

If the number of switches is large and exceeds the number of bits of the system data bus, the switch is usually arranged in a matrix form to simplify the I/O interface circuit. The keyboard in the computer is a typical example. Figure 5-26 for the matrix keyboard interface circuit, the data bus $D_7 \sim D_0$ through the interface of the data output latch to drive the keyboard matrix 8 lines $R_7 \sim R_0$, keyboard matrix 8 column line $C_7 \sim C_0$. The information can be sent to the data buses D_7 to D_0 via the data input buffer. The control signals of the two ports are generated by the address decoding circuit.

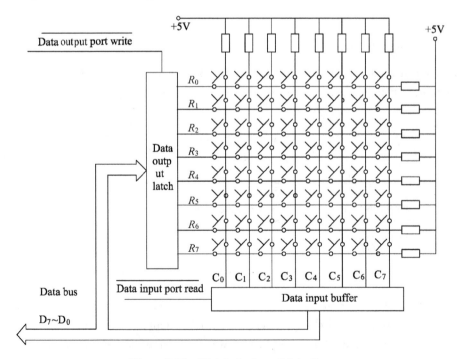

Figure 5-26 Matrix keyboard interface

182 / assembly LANGUAGE & Interfacing Technology

When a key in the keyboard matrix is pressed, the line and column lines to which it is connected are turned on. Obviously:

(1) If no key is pressed, the read column port gets all "1".

(2) When "1" is output from the line port, the read column port is all "1" regardless of whether or not the key is pressed.

(3) When the output of a bit line is "0", if the key is pressed on the line, the column line of the key is "0" and the column of the remaining columns is "1".

Each key in the matrix has a unique 2-byte "row and column". For example, the intersection of the line R_3 and the column line C_2 is "F7FBH", the high byte is the row information, F7H =11110111B, "0" appears in the D_3 bit, and the data line D_3 is connected to the line through the line R_3 indicates that the key is in the R_3 line; the low byte is the column information, FBH =11111011B, "0" appears in the D_2 bit, and the data line D_2 is connected to the column line C_2 through the column port, indicating that the key is in the C_2 column.

To identify a single key pressed in the keyboard matrix by executing a program, the program flow is:

(1) Output all "0" to the line port and read the column port. If the column port information is full "1" then exit, otherwise perform step 2; (this operation is called "rough scan").

(2) Delay shaking, usually 20ms delay.

(3) Read the column port again, if the read information is full "1", that no key to press, exit, otherwise the next step.

(4) Output "0" line by line and read the column port. Until the read column information is not all "1" (1 bit is 0), identify the key, perform step 5, otherwise exit; (this operation is called "fine scan").

(5) The formation of the key 2-byte bar code, the current line output value for the key line where the information, read the value of the column where the key information.

The method of identifying a single key that is pressed in the keyboard matrix is called a "line scan" above.

The following subroutine scans the keyboard matrix based on the line scan method. When the key is pressed, it returns the 2-byte bar code of the key through the AX, AH is the line code, AL is the column code, and no key returns "−1". R_PORT, C_PORT, respectively, for the ranks of the port address, by the address decoding circuit to determine the specific value.

```
SCANKEY PROC
        PUSH CX
        PUSH DX
        MOV DX, R_PORT
        MOV AL, 0
        OUT DX, AL
        MOV DX, C_PORT
        IN AL,DX
```

```
                CMP AL,0FFH
                JZ NO_KEY
                ; delay 20 MS
                IN AL,DX
                CMP AL,0FFH
                JZ NO_KEY
                MOV AH,0FEH
                MOV CX,8
        NEXT:   MOV AL,AH
                ROL AH,1
                MOV DX,R_PORT
                OUT DX,AL
                MOV DX,C_PORT
                IN AL,DX
                CMP AL,0FFH
                LOOPZ NEXT
                JZ NO_KEY
                ROR AH,1
                JMP EXIT
        NO_KEY: MOV AX,-1
        EXIT:   POP DX
                POP CX
                RET
        SCANKEY ENDP
```

Get the key "bar code", you can use it to look up the table, and further get its actual code (such as key information ASCII code).

For example, if the 64 keys in Figure 5-26 are from top to bottom, from left to right are defined as 0~9, A~Z, a~z, carriage return and line feed, the following program is scanned by calling the keyboard. The subroutine recognizes the key and displays the actual information of the key on the screen.

```
        .MODEL SMALL
        .DATA
            KEYVALUE DB '01..9A..Za..z',0DH,0AH     ; ASCII code for 64 keys
            KEYCODE DW 0FEFEH,0FEFDH,...,7F7FH      ; Columns of 64 keys
            R_PORT DW ?
            C_PORT DW ?
        .STACK 100H
        .CODE
        START:  MOV AX,@DATA
                MOV DS,AX
                CALL SCANKEY
                CMP AX,-1
                JZ QUIT
                MOV CX,64
                LEA SI,KEYVALUE-1
```

```
                 LEA DI,KEYCODE - 2
SCANTAB: INC SI
         INC DI
         INC DI
         CMP AX, [DI]
      LOOPNZ SCANTAB
         JNZ QUIT
         MOV DL,[SI]
         MOV AH,2
         INT 21H
   QUIT: MOV AX,4C00H
         INT 21H
             END START
```

In summary, the input switch into the form of array, which can be a small hardware cost to access more than the system data bus bit switch, but it the need to prepare a more complex procedures to identify the status of each switch.

Exercises 5

5.1 In the computer system, why do peripherals need to connect to the host through the I/O interface? What are the I/O interfaces?

5.2 Which function modules should I normally include in the I/O interface? According to the stored information which is different, which port is divided into?

5.3 What is the address of the port? Based on the Intel series of microprocessors in the computer system, which is the way of the port address arrangement? What is the number of ports that can be accessed? What instructions are used to access the port?

5.4 What is the port's "input buffer" and "output latch"? What are the simple input ports and output ports that usually consist of?

5.5 8088 What is the function of the instruction "IN AL, DX" in the smallest computer system? If (DX) =120H, what type of bus operation is required during the execution of the instruction? What is the information output on the 8088 related pin in each T state of the bus cycle?

5.6 Host and peripherals to transfer information between the control of what kind of? What are the characteristics of each? Briefly describe the workflow of the program query mode.

5.7 Read the query input interface circuit in Figure 5-19 and write the assembly language program that completes the following functions: Read 100 bytes of data from the input device in query mode and store it in the data buffer memory INBUFF.

5.8 Read the query output interface circuit in Figure 5-20 and write an assembly language program that performs the following functions: A number of bytes of data defined in the data segment memory buffer OUTBUFF are sent to the output device in a query mode, When the data of the area is "0DH", the program ends after the data is output.

5.9 Read the Figure 5-22 switch output interface, if the digital tube for the common cathode, and the system data lines $D_7 \sim D_0$ from high to low order to connect 74LS273 the

$D_8 \sim D_1$ pin, the $0 \sim 9$ ten-shaped information word code What is it? If the port address is 310H, write in the digital display "2" assembly language program fragment.

5.10 Introduction the complete process of using the "line scan method" to identify the keys pressed in the keyboard matrix. Figure 5-29 Matrix keyboard interface, respectively, to determine (R_0, C_5), (R_2, C_7), (R_6, C_6) at the button 2 bytes.

CHAPTER 6
INTERRUPTION SYSTEM FOR MICROCOMPUTERS

【Abstract】 Interruption in is an important function that modern computer, but also the development of the history of computer is an important milepost. This chapter first introduces the basic concepts of interrupt, then tells the 8086 interrupt function of the system, and then the 8259 interrupt controller the principle and programming, finally the method of interrupt program design.

【Learning Goal】
· Grasp the basic concepts of interrupts, interrupts, classifications, etc.
· Master the method of accessing the address of the interrupt service program under the real address mode.
· Understanding the interrupt response process for interrupts and masking interrupts.
· Master the working principle of programmable logic controller 8259A.
· Understand the working style of 8259A, and master the programming method and application of 8259A.
· Master interrupt programming and applications.

6.1 The basic concept of interrupt system

6.1.1 The basic concept of interruption

The interruption is an important technology in the computer, and is a major way to exchange information processor and external devices in modern computers. The modern system of microcomputer keyboard, mouse and other I/O interface are in the interrupt mode and the processor for the exchange of information. What is the interruption? One in life example: Mike is reading his book when his mobile phone suddenly rang. At this time, Mike closes the book and answers the phone, when the call completed, Mike goes back reading. This example shows that the interruption and the treatment process: mobile phone ring makes Mike abort the current work temporarily, and deal with sudden or need to deal with things (answer the phone and deal with the things that need to be dealt with in real time, and then proceed with the original). This process is called an interrupt, as shown in Figure 6-1.

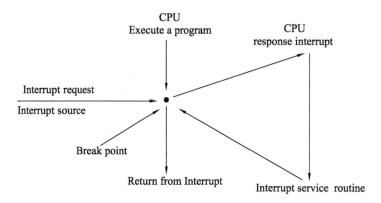

Figure 6-1   Interrupt process

In this process, the interruption caused by the internal or external events called interrupt sources. In order to distinguish between different interrupt sources, CPU for each interrupt source is assigned a unique number, called the interrupt type number (also known as the interrupt vector number). The interrupt source to the request of CPU is called an interrupt request. Because of the disruption, the forthcoming implementation of program execution is suspended, but the instruction has not been executed the address is called a breakpoint, i. e. The interrupt occurs the CS and IP values. The interrupt event service program is called the interrupt service program. The different interrupt sources corresponding to different when the interrupt service routine. Interruption and response should be submitted, the interrupt source interrupt type number it to CPU (whether it is submitted in what way), according to the type of CPU, find the corresponding interrupt service program entrance address. Some microcomputer interrupt service program is compiled and stored in the ROM, or by the user according to the needs of developing the use of.

The process is similar to the calling process interrupt subroutine, there are procedures of the switch, but there is an essential difference between the subroutine execution call by the main program, therefore is determined. The interrupt service routine caused by an event, it is random and uncertain. Of course, it can also be used to achieve a specific function call instruction INT, but with the call processing mechanism subroutine is different.

CPU receives the interrupt request, and the process from the current program to the interruption of the service routine is called the interrupt response. When the interrupt service ends, the program is returned to the interrupted program. The process of continuing the execution is called interrupt return

6.1.2   Interrupt the function of the system

An interrupt system is a collection of software and hardware that implements interrupt functions. Interrupt systems should have the following functions in order to meet the interruption requests in various situations:

(1) To achieve interrupt response and interrupt, the interrupt service returns. When an interrupt source sends an interrupt request, CPU can decide whether to respond to the interrupt request. If allowed to respond to the interrupt request, CPU can protect the breakpoint, transfers control to the interrupt service program corresponding to the interrupt processing,

CPU returns to the breakpoint and continues to implement the original program.

(2) To achieve the priority queue interrupt. When there are two or more interrupt source, interrupt request is put forward at the same time, interrupt system according to the nature of the interrupt source to distinguish the order of priority, priority is given to ensure the high priority interrupt request.

(3) To achieve interrupt nesting in the interrupt handling process. If there is a new high priority interrupt, interrupt system to enable CPU to suspend the current interrupt service program execution, and to respond to and deal with higher priority interrupt, the interrupt service program after the end of treatment and return to the original priority level is low. This situation is called nested interrupt or multiple interrupt.

After the computer system has the perfect interruption function, the overall performance of the system can be greatly improved, mainly reflected in the following aspects:

(1) The ability of parallel processing. The interrupt system, CPU can also work with a number of peripherals CPU peripheral equipment after initialization, you can run other programs. When CPU and peripherals need each other to exchange information, "break" the current CPU. This CPU can simultaneously control a plurality of peripheral parallel work greatly to improve the system throughput and efficiency.

(2) Real time processing ability. When applied to real-time control, many of the information on the scene needs CPU to respond quickly and deal with in time, and the request time is often random. Only interrupt system can realize real-time processing

(3) The fault processing ability. In the operation process of CPU, tend to be some trouble. For example, a power failure (the voltage drop is too large, 220V 160V to continue to drop), memory read and write error, calculation error, can use the interrupt system function, automatically turn to perform troubleshooting procedures, and the other does not affect the execution of the program.

(4) Multi task operation. Through timing interruption, under the scheduling of operating system, the time of CPU can be assigned to multiple tasks, and the tasks run alternately. From a macro point of view, multiple tasks are running at the same time

6.1.3 Interrupt handling process

1. Interrupt request

The interrupt request is the first step in the interrupt process. The interrupt generator's interrupt request condition is caused by an interrupt source. The internal interrupt is caused by a specific flag or instruction of the CPU. The external interrupt is an interrupt request signal generated by interrupting the interface circuit. In the input mode, for example, interrupt interface circuit shown in Figure 6-2. After the external device is ready for data, issue the $\overline{STB}$ signal to insert the data into the latch and set the interrupt request trigger. If the request is not masked, an interrupt request signal is generated. Take the 8086 processor as an example, and the interrupt request signal is sent to the INTR pin of the CPU. The CPU sends the $\overline{INTA}$ response signal in response to the request to obtain the interrupt type number.

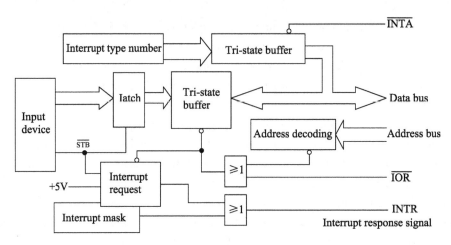

Figure 6-2 Interrupt input interface

2. Interrupt priority

Because the type of interrupt sources are generated and external interrupt is random, there may be two or more than two interrupt sources and puts forward the interrupt request. It must be according to the order of priority interrupt source, interrupt source to determine an interrupt priority for each level. When multiple interrupt sources interrupt requests at the same time, interrupt system identifies the highest priority request, and first response to the request, and then respond to lower level request. There are two methods for setting interrupt priority: software query priority queue and hardware priority discrimination which includes daisy chain and vector queue priority.

(1) Software query priority queuing

After receiving the interrupt request, CPU executes the query program, reading the interrupt request register state, in accordance with the order by a query, when an interrupt request state is effective, it will be transferred to the corresponding interrupt service subroutine. First query interrupt source is high priority, later is a lower priority.

Software query priority queue has the advantages of simple interface circuit, and the priority order with the query sequence change, modify convenient; the disadvantage is the interrupt source more, by the interrupt to enter the interrupt service program is interrupted for long periods of time, slow response, this method not only in the interrupt source, real-time less demanding occasions.

(2) Chain priority queuing method daisy chain method

A priority logic circuit is connected to each corresponding peripheral interface circuit, the logic circuit to form a chain, called daisy chain. As shown in Figure 6-3, controlled by the daisy chain interrupt signal pathway, so as to control the equipment priority. Any device can send an interrupt request to CPU to apply a interrupt. CPU answers this signal and sends $\overline{INTA}$. In the daisy chain, if the interface no interrupt request, the signal will be transmit backward; if an interface is appeared, signal $\overline{INTA}$ will be blocked, not backward transfer. Through the logic circuit, signals $\overline{INTA}$ from the near port of CPU, along the daisy chain step by step backward transfer, until it is a block interface interrupt requests so far. There is a plurality of interrupt sources and requests, the most close to the CPU interface first to get the response of CPU, the highest priority, the farther from the processor interface, the lower priority.

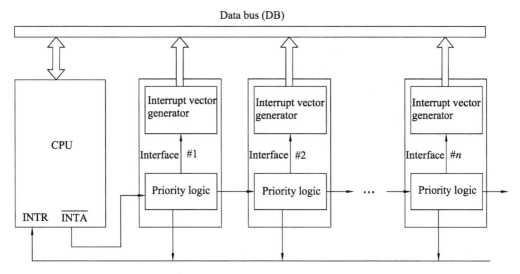

Figure 6-3 Chain priority queuing circuit

(3) Interrupt controller-vector priority queuing special circuit

At present, the most widely used in microcomputer system programmable interrupt controller, is responsible for receiving the interrupt requests on shielding, arbitration. It can receive multiple I/O interface from the interrupt request signal an interrupt controller, a priority by priority arbiter management. Priority arbiter in hardware have on the priority order of each request to do the arrangements, but also can set different priority through the procedure, to dynamically modify the priority objective. The interrupt controller interrupt request to the CPU arbitration after the interrupt request signal INTR, CPU echo signal response, and then the interrupt controller sends the interrupt type number (also called vector number) to the CPU, and CPU finds the corresponding interrupt service subroutine according to the class type and executes it.

3. Interrupt response

The interrupt request received from the CPU, to turn to implement the interrupt service program is interrupted in response CPU after the end of each instruction to the sampling interrupt request signal, if the detected interrupt request, and in response to allow interruption, the system automatically enter the response cycle is completed by hardware interrupt, save breakpoint take the entrance address, interrupt service program and a series of interrupt response operations.

4. Interrupt handing

Interrupt processing is the execution of an interrupt service routine that completes the operation required by the interrupt source. The interrupt handler usually consists of the following sections:

(1) Protection of the scene. Generally, the register contents used in the interrupt service program are pressed into the stack through the stack instruction. This is called the protection

site, so that the original program can be executed correctly after the interrupt service program returns.

(2) Perform the interrupt service program, which is the core of the interrupt processing and completes the task required by the interrupt source.

(3) Restore the field. Use stack instructions to restore the contents of the registers concerned with the protection of the scene and to ensure that the stack pointer is restored to the point when it enters the interrupt processing.

5. Return from interrupt

By interrupting the return instruction, messages such as breakpoints and flags stored in the stack are ejected and returned to where the original program has been interrupted.

6.2 8086 CPU interrupt system

6.2.1 Classification of 8086 interrupts

8086 uses vector interrupt mechanism, using 8-bit binary code to distinguish between different interrupt sources, and it can support 256 different interrupts. Depending on the cause and location of the interrupt, it is divided into internal interrupts and external interrupts. The external interrupts are divided into maskable interrupts and non-maskable interrupts, as shown in Figure 6-4.

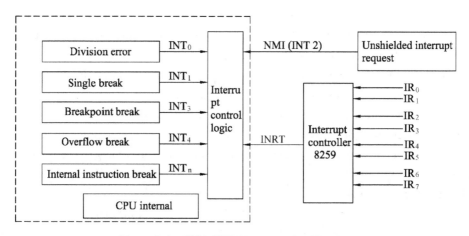

Figure 6-4 8086 CPU interrupt classification

1. External interrupt

CPU external interrupt is due to external interrupt request of the proposed program interrupt, also known as CPU. The hardware interrupt triggered by external pins, divided into maskable interrupt and non maskable interrupt. What can be shielded and unshielded? Here is demonstrated by an example in life: when the students are in class, they need to concentrate on class, in the break state, if the students have mobile phone calls, cannot answer the phone (because the school is closed for the interrupt status, answering the phone to be shielded, so cannot response to the interruption). But if the call is in the class, the students can answer the phone (off). It is the same class. When an earthquake or fire emergency is serious, the students must handle the event immediately (because of the earthquake, the fire emergency is not blocked, as long as the unshielded events must be processed).

(1) Unshielded interrupt NMI. 8086 introduced by the NMI pin, the rising edge triggered. When CPU sampling from low to high the rising edge of the signal after the NMI service program to automatically enter the NMI pin, it is not affected by the interrupt flag IF limit, regardless of the current IF =1 or IF =0 CPU will respond to the request of. NMI the interrupt is mainly used to deal with external emergency events, such as power failure, memory error. Interrupt type number is 2.

(2) Maskable interrupt INTR. 8086 maskable interrupt introduced by the INTR pin, high effective. CPU at the end of a clock cycle the current instruction cycle sampling INTR pin, therefore, the INTR request signal must be maintained at least to the current instruction is executed before the end of may be a response. If there is a maskable interrupt request, according to the CPU interrupt flag state determines whether the response of IF. When IF =1, CPU in response to a INTR request; ban IF =0. In response to allow maskable interrupt under the condition of CPU in the current instruction execution output after the maskable interrupt respond to a response signal $\overline{INTA}$.

CPU is only an INTR pin, how to manage multiple external interrupt request, how to obtain the interrupt type number in each interrupt source? 8086/8088 maskable interrupt usually use a dedicated interrupt controller Intel 8259A to achieve unified management. Maskable interrupt, interrupt detail type number supplied by the. 8259A interrupt controller in Section 6.3.

The shielded interrupt has the following characteristics:

(1) Interrupts are caused by external events and require hardware circuitry to generate interrupt request signals.

(2) The interrupt request can be interrupted by the response, allowing for the impact of flag IF.

(3) When responding to a request, a response $\overline{INTA}$ is sent to the 8259A via the signal, and the interrupt type number is provided by the 8259.

(4) Allow interrupt nesting.

(5) A masked request has randomness, and the timing of the occurrence is unpredictable.

2. Internal interrupt

The internal interrupt is also known as the software interrupt, an interrupt is detected by the CPU abnormality or execution of the software interrupt instruction caused, belonging to the non maskable interrupt. Internal interrupt is generally in the execution of a command or a flag register according to the flag state and the general division error, step interruption, breakpoint interrupt, overflow INT interrupt and N interrupt.

(1) Division interrupts. When division instructions (DIV or IDIV) are executed, if the divisor is found to be 0 or the quotient exceeds the range represented by the register, the 8086 CPU immediately executes an internal interrupt with an interrupt type number of 0.

(2) Single-step interrupt. If the self-trapping flag TF =1 of the flag register in the 8086 CPU, the CPU executes an internal interrupt each time an instruction is executed. It is used to achieve the program single-step debugging, such as DEBUG under the implementation of the T

command is achieved through a single-step interrupt.

(3) Breakpoint break. The instruction INT 3 generates an internal interrupt with an interrupt type of 3, called a breakpoint interrupt. In the process of debugging procedures, the need to track the program to understand the process of intermediate results, you can use the INT 3 command to temporarily replace the original instruction, called set breakpoints. Program execution to the breakpoint, will be due to the implementation of INT 3 instructions into the type 3 interrupt service routine. As a result, the original procedure was suspended. At this point you can read the program execution environment (instruction address, register value, variable value) for the programmer to use debugging. Finally restore the original instruction, continue to implement the program being debugged.

(4) Overflow interrupt. 8086 flag register overflow flag OF =1, then execute the INTO instruction to produce an internal interrupt of type 4; otherwise, this command does not work, the program sequentially executes the next instruction.

(5) Instruction execution interruption. INT N interrupt (N interrupt type number, ranging from 0 to 255) generated by the type number N interrupt. In the program, the interrupt service program users can use the INT n command to easily call different interrupt type number.

To sum up, software interrupts have the following characteristics:

(1) Interrupts are caused by CPU internal reasons and are independent of external circuitry. The interrupt type number is automatically supplied by the CPU.

(2) In addition to single step interrupts, internal interrupts cannot be prohibited by software and are not affected by the IF state.

(3) Internal interrupts are random, similar to call subroutines.

3. Priority of interruption

8086 CPU interrupt priority from high to low order is: internal interrupt (except for single step interrupt), >NMI interrupt, >INTR interrupt > single step interrupt. Except for single step interrupt, any internal interrupt priority is higher than external interrupt.

6.2.2 Interrupt vector table

1. Interrupt vector

The interrupt vector refers to the entrance address of interrupt service program, including the base address and the offset address. Each interrupt vector is 4 bytes, the low two bytes, are offset address, the high two bytes are segment address.

2. Interrupt vector table

8086 system from the memory 00000H ~ 0003FFH a total of 1024 bytes by the interrupt type from low to high order followed by 256 interrupt source interrupt vector, called the interrupt vector table, as shown in Figure 6-5.

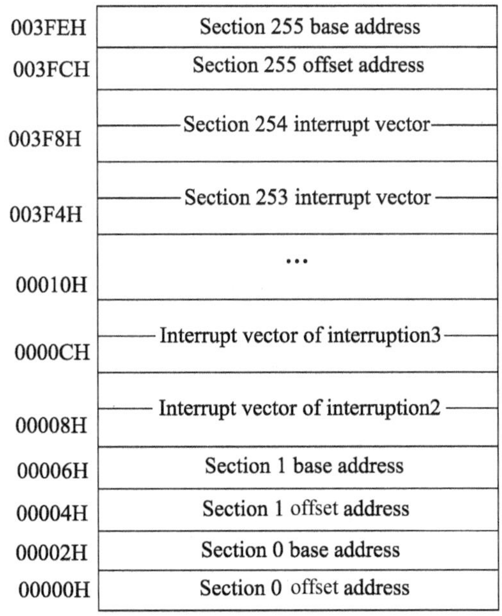

Figure 6-5 8086 interrupt vector table

```
003FEH  | Section 255 base address
003FCH  | Section 255 offset address
003F8H  |— Section 254 interrupt vector —
003F4H  |— Section 253 interrupt vector —
00010H  |          ...
0000CH  |— Interrupt vector of interruption3 —
00008H  |— Interrupt vector of interruption2 —
00006H  | Section 1 base address
00004H  | Section 1 offset address
00002H  | Section 0 base address
00000H  | Section 0 offset address
```

Figure 6-5 8086 interrupt vector table

The position of the interrupt vector in the table is called the interrupt vector address, and the relation between the interrupt vector address and the interrupt type number is:

Interrupt vector address = interrupt type number * 4

After getting interrupt type number, according to the interrupt vector address corresponding to the interrupt vector in the interrupt vector table in order to find and execute the interrupt service program. For example, interrupt type number 18H interrupt vector 1234H:5600H, the vector interrupt vector address for 18H * 4 = 60H. From 60H to 63H so the memory of these 4 units the contents are as follows: 00H, 56H, 34H, 12H.

8086 can manage 256 interrupts, as shown in Table 6-1, 00H ~ 04H for CPU-specific interrupt, 05H ~ 3FH for the system to retain the interrupt, 40H ~ FFH for the user-defined interrupt. System interruptions include: 08H ~ 0FH hardware interrupt; 10H ~ 1FH BIOS call and 20H ~ 3FH DOS system call, these interrupt users generally cannot change their definition.

Table 6-1 8086 classification of interrupt vector tables

00H ~ 04H——System specific	10H ~ 1FH——BIOS uses	40H ~ FFH——user uses
08H ~ 0FH——Hardware interrupt	20H ~ 3FH——DOS uses	

The interrupt vector table DOS, BIOS soft interrupt and system dedicated interrupt vector, when the system is initialized. Fashion into the interrupt service program developed by users before use, must be in the interrupt vector table set the corresponding interrupt vector.

3. Method for setting interrupt vector

Suppose the interrupt type number is N, and the interrupt service routine is called INTRPROC. The following two methods can complete the setting of the interrupt vector

(1) Use the 25H DOS function call to set the interrupt vector. The program segment is as follows:

```
        MOV AX,SEG INTRPROC
        MOV, DS, AX        ; the entry address of the preset interrupt service
                             program in DS:DX
    MOV DX,OFFSET INTRPROC
    MOV, AH, 25H           ; call function number for DOS in AH
    MOV, AL, N             ; preset interrupt type number in AL
    INT 21H                ; DOS call
```

(2) Using the MOV instruction to set the interrupt vector:

```
    MOV AX,0
        MOV ES, AX         ; the interrupt vector table segment base address is
                             0000H
    MOV, BX, N*4           ; preset interrupt vector address in BX
    MOV AX,OFFSET INTRPROC
    MOV, ES:[BX], AX       ; the address of the vector table is low and the offset
                             address is stored
    MOV AX,SEG INTRPROC
    MOV, ES:[BX+2], AX;    base address of high address storage in vector
```

6.2.3 Response of 8086 to interrupt

CPU response interrupts must satisfy the following conditions:

(1) CPU receives the interrupt request.

(2) No DMA request.

(3) The end of the current instruction execution.

(4) If it is an INTR interrupt request, the CPU must also open interrupts, that is, the interrupt flag IF =1.

The response condition, enter the interrupt response time and interrupt response operation automatically by the hardware. 8086 CPU on different types of interrupt response process is slightly different, which is the main process to obtain the number of different types of interruption, after obtaining the interrupt type number, CPU is the same as the response process of them.

NMI and internal interrupt, interrupt type number is known. For example, NMI interrupt type number for 2, single step interruption, other types of fixed overflow interrupt is 1 and 4, the type of N, INT interrupt instruction by instruction given directly. Maskable interrupt INTR interrupt type by external interrupt control interface (usually interrupt controller 8259A). For the INTR interrupt, the interrupt response into the cycle, first get the interrupt type number, as follows:

(1) CPU interrupts the response signal of a negative pulse from the pin $\overline{INTA}$ and informs the external interrupt control interface. The interrupt has been answered. The interrupt control interface sends the interrupt type number to the system data bus.

(2) Send out second response signals $\overline{INTA}$ to get the interrupt type number.

After obtaining the interrupt type, 8086 responds to the same response as the different interrupts, as follows:

(1) Press the value of the flag register FLAGS into the stack. At the same time, the

interrupt enable flag IF in FLAGS and the single step flag TF are cleared 0 to shield other external interrupt requests, and CPU is prevented from executing the interrupt service program in a single step.

(2) Protect breakpoint, first press the value of the current code segment register CS into the stack, then put the value of the instruction pointer IP into the stack, so that after the interrupt processing is completed, it can return to the main program correctly and continue to execute.

(3) It will interrupt the type number left two bits (i.e. multiplied by 4), the formation of the interrupt vector address, according to the interrupt vector address to the interrupt vector table to find the interrupt vector, the offset address will interrupt service program written in IP; segment base written CS. control transferred to the interrupt service program execution.

6.3 Interrupt controller 8259A

8259A is a programmable interrupt control chip produced by Intel Corporation. It has powerful interrupt management function, and the main functions are as follows:

(1) It has 8 level priority control, and can be extended to 64 level priority control by cascading.

(2) Each level of interruption may be shielded or allowed by the program separately.

(3) The interrupt type number can be sent to CPU.

(4) You can choose a number of different ways of programming by programming so that you can adapt to various system requirements.

6.3.1 Pin signal of 8259A

8259A is the CMOS process interface chip, DIP package, a total of 28 pins, as shown in Figure 6-6.

1. Pin signal connected to the CPU

$\overline{CS}$: The chip select signal, input, active low. Effective, CPU can read and write to 8259A.

$\overline{RD}$: Read signal, input, active low, both $\overline{CS}$ and $\overline{RD}$ are valid. Allow the CPU to read the status signal of the 8259A.

$\overline{WR}$: Write signal, input, active low, both $\overline{CS}$ and $\overline{WR}$ are valid. Allow 8259A to receive command word sent by CPU.

A0: address line, enter. Select the two programmable ports within the 8259A.

The $D_0 \sim D_7$: 8-bit bidirectional data bus is used to transmit control command words, status, and interrupt type numbers.

INT: interrupt request signal, the output. 8259A sends the interrupt request to the CPU through this signal.

$\overline{INTA}$: The interrupt response signal, input, active low, receives the interrupt response signal $\overline{INTA}$ from the CPU.

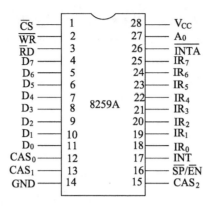

Figure 6-6 8259A pin diagram

2. Pin signal that is connected to the interrupt source

$IR_0 \sim IR_7$: external interrupt request signal, input) active at high or rising edge. Receive external interrupt request signal.

3. Pin signal during cascading expansion

$CAS_0 \sim CAS_2$: cascaded signals, bidirectional. 8259A work in cascade mode, master chip, $CAS_0 \sim CAS_2$ for output, 8259A works from chip, and $CAS_0 \sim CAS_2$ as input.

$\overline{SP/EN}$: From/buffer allows the control signal. When the 8259A is in a non buffer mode, this pin is the input, $\overline{SP} = 0$ shows that the 8259A is slowe chip, $\overline{SP} = 1$ shows the 8259A is master chip. When the 8259A is in the buffer mode, 8259A buffer is connected by bus and data bus, the output signal $\overline{EN}$ to control the bus buffer.

6.3.2 The internal structure of the 8259A

The internal structure of the 8259A is shown in Figure 6-7, which consists of the following components.

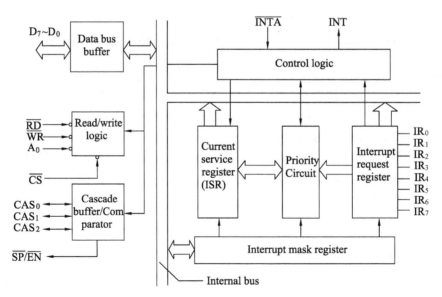

Figure 6-7 8259A internal structure diagram

CHAPTER 6 INTERRUPTION SYSTEM FOR MICROCOMPUTERS

1. Interrupt request register IRR

Interrupt request register is an 8-bit register latch function, stored external input interrupt request signal $IR_0 \sim IR_7$. When IR_i ($i = 0 \sim 7$) end interrupt request, the corresponding IRR IRR_i is set to "1". IRR_i bit will be cleared when $\overline{INTA}$ is valid, the set and clear of IRR are compated by hardware.

2. Interrupt mask register IMR

Interrupt mask register is an 8-bit register, shielding information used to store the external request corresponding to the IR0 ~ IR7. When IMR I is "0", enter the interrupt request interrupt priority arbiter IRR_i bit in the IRR register is allowed. When IMR I is "1", the IRR_i interrupt request is blocked, the prohibition to interrupt priority arbiter in the IMR set and cleared by software programming.

3. Interrupt service register ISR

Interrupt service register is the 8-bit register for storing is interrupted with the status of the 8259A services, when ISR I is "1" bit interrupt IRR_i in service; "0" means there is no service. ISR I "1" CPU is currently the service interruption may be interrupted, it may be not yet completed service interrupted nested pending interrupt in. ISR I interrupt service after the end can be cleared by software programming.

4. Interrupt priority discriminator PR

Interrupt priority arbiter, namely priority arbitration logic, it can be external interrupt request priority queue, IMR interrupt priority interruption in service request and through ISR, if a new interrupt request service than ISR in state of high priority, will apply to CPU interrupt the operation process. All completed by hardware, the register is not accessible.

5. Data bus buffer

The data bus buffer is a two-way three state 8-bit buffer, is the interface between 8259A and system data bus. 8259A uses it to receive control word from CPU, as well as send interrupt type and its status to CPU.

6. Control logic

The control module of internal 8259A, each part of the chip internal work in accordance with the provisions of everything in good order and well arranged program. With 7-8-bit programmable registers, including 4 initialization command register ICW (Initialization Command Word) and 3 OCW (Operation Command operation command word Word). The computer system initialization command word is started by set the initialization procedure, initialization command word once set, generally in the working process of the system is no longer changed. Operation command word is set by the application is used to perform dynamic control of interrupt processing. The command word can be set many times during system operation.

7. Read-write control logic

CPU finishes the initialization of 8259A programming and fetching its status through $\overline{CS}$, $AO, \overline{RD}$ and $\overline{WR}$ four signals.

Output instruction makes $\overline{WR}$ valid and writes 8259A command word to ICW or OCW valid data bus when cpu writes 8259A. IN instruction makes $\overline{RD}$ valid and CPU when CPU reads 8259A reads contents from IRR, ISR or IMP viadata bus.

8. Cascaded Buffers/Comparators

The cascade between the 8259A chips, determine the master chip and slave chip. Associated with this component has three cascaded line $CAS_0 \sim CAS_2$ and a master set/buffer read and write control signal $\overline{SP/EN}$.

If the system is used in multi 8259A cascade system, in order to reduce the burden of the system's data bus, the data can be collected after each 8259A bus through a bidirectional buffer connected to the system data bus. This way is called a buffer made. This pin $\overline{SP/EN}$ is used as $\overline{EN}$ control bus buffer open. When the system is only a few 8259A, does not work in the buffer mode, $\overline{SP/EN}$ used as $\overline{SR}$, high level, indicates that 8259A is master, low level, indicates that the 8259A is slave.

6.3.3 Working process of 8259A

8259A a complete working process is as follows:

(1) The interrupt source raises the interrupt request at the interrupt request input $IR_0 \sim IR_7$.

(2) The interrupt request is latched in the IRR register, and the IMR shield, the results give priority decision circuit and ISR registers. The sub optimal interrupt request priority is higher than the service priority, the controller receives an interrupt request signal sent to CPU INT.

(3) CPU receives the 8259A request signal from the INTR pin, and if the IF is 1, allowing interrupts, two consecutive $\overline{INTA}$ signals are sent to respond to the interrupt.

(4) 8259A receives the first signal, $\overline{INTA}$ which corresponds to the IRR bit 0 of the interrupt source and the ISR corresponding position 1, thus prohibiting peer and low-level interrupts.

(5) If 8259A is used as the main control interrupt controller, it sends the cascaded address from $CAS_0 \sim CAS_2$ during the fist cycle of $\overline{INTA}$. If 8259A is used alone or slave controller chosen by $CAS_0 \sim CAS_2$, it sends interrupt during the second cycle of $\overline{INTA}$. type number from $D_0 \sim D_7$.

(6) If 8259A is set to automatic termination, then the corresponding bits of ISR are automatically cleared 0 after the second $\overline{INTA}$.

(7) CPU reads the type number and goes to the corresponding interrupt handler

(8) Before the end of the interrupt processing, if the 8259A is set to the end of the not automatic interrupt, a EOI (end of interrupt) command is sent by the interrupt handler, so that the corresponding bits of the ISR are cleared. Interrupt is over.

6.3.4 How does the 8259A work

1. The way the interrupt request is triggered

The interrupt request trigger mode is the valid signal type of the interrupt request pin IR0 ~ IR7, in two ways:

(1) Edge triggered mode

In the edge triggered mode, the 8259A interrupts the rising edge of the requested input as the interrupt request signal, and the interrupt request input can always remain high after the rising edge triggered signal appears.

(2) Level trigger mode

In level triggered mode, 8259A will interrupt high level request input appears as an interrupt request signal. But when the interrupt response and interrupt input must promptly cancel the high level, or in the CPU interrupt handling process, and without interruption, the high level of original input can cause second interrupt wrong.

The LTIM bits in the initialization command word ICW_1 are used to set the two trigger modes, LTIM =1, to set the level trigger mode, and LTIM =0 to the edge triggered mode.

2. Priority management

8259A has two methods of determining priorities: fixed priority and loop priority. Each type has different implementations.

(1) Fixed priority. This way, each pin numbers interrupt source priority connected by its decision, the fixed order of priority: $IR_0 > IR_1 > IR_7$. Interrupt source once connected, priority is determined. According to the different ways of nesting is also divided into fully nested mode and special fully nested mode.

Full nested mode is the most common way for 8259A. After initialization, the system defaults to working in a fully nested manner. In this way, priority is nested only if the priority is higher than the interrupt request that is being serviced.

Special fully nested mode can achieve the advanced interrupt request nesting, it is set to 8259A. When there are multiple cascade 8259A system, a main piece, the other from the film. From the hypothesis set to fully nested mode, when an interrupt request from entering and processing is the same, from on chip has a higher priority interrupt request, from the response to the request, and to the main application interruption, but the main piece is the interrupt request. If the main piece of the nested manner does not response from the same pin at the request. When the main plate is in a special full nested working way, allow the same level of master interrupt Nested requests.

Therefore, when the system has only one slice of 8259A, it usually adopts a full nested method. When there are multiple 8259A in the system, the main slice must work in a special full nested manner, and the whole nested mode is adopted from the chip.

(2) Cyclic priority. The priority of each interrupt source can be changed dynamically. There are two kinds of priority automatic cycle mode and priority special cycle mode.

Priority automatic cycle mode, the initial order of priority and fixed priority, IR_0 is the highest, the lowest IR_7. But when a certain level interrupt response, the interrupt priority level automatically reduced to a minimum. For example, 8259A work in priority automatic cycle mode, IR_4 request response, interrupt priority from high to low. The order is: IR_5, IR_6, IR_7, IR_0, IR_1, IR_2, IR_3, IR_4.

The initial order of priority in special priority loops is not $IR_0 > IR_1 > IR_7$, the initial minimum priority is determined by programming, and the rest is similar to the automatic loop method. The minimum priority is specified by the operation command word OCW_2.

3. How to block the interrupt source

CPU by the CLI directive to ban all in response to interrupt, and interrupt 8259A refers to the peripheral interrupt application, which allows or does not allow peripherals to the interruption of CPU applications instead of. 8259A interrupt response ring problem has been proposed two kinds of interrupt mode:

(1) General mask mode

The interrupt mask register IMR in one or several position "1", can be a corresponding interrupt request masked. The corresponding IR pin even if an interrupt request cannot be sent to the priority decision. "0" open the corresponding interrupt.

The settings for IMR are usually in the main program, but they can also be placed in the interrupt service program, depending on the interrupt handling requirements

(2) Special mask mode

Special mask methods, a IMR_i is set to "1", the corresponding bits of ISR will also be cleared "0". On certain occasions, we hope to dynamically change the structure of the system a priority interrupt service program. For example, when the CPU is in the process of the interrupt program, at the same time, CPU wants to open the lower interrupt request, then using a special way to mask the interrupt and allow the priority high or low interrupt entry. Special way is always used in the interrupt handler. For example, the currently executing IR_3 interrupt mask service program, set up a special mask after using the OCW_1 to interrupt mask register IMR_3 position "1", will also make ISR_3 interrupt service register in the automatic clear "0", such as mask and currently being processed by the same level of interruption, and open the interruption of the lower level. This method is rarely used.

4. End of interrupt mode

End of interrupt is the essence of clear all ISR bits of "1", it means to cancel the corresponding interrupt level, so that the low priority interrupt source can apply for interrupt. If the service is completed, the "1" bit is cleared to "0", if not, some interrupts will alway occupy the interrupt level, then the low level of the interrupt request cannot be provided. 8259A has automatic and not automatic two ways.

(1) Automatic ending mode (AEOI)

AEOI refers to the interrupt response time, automatically clear the ISR register to "0". Therefore, in the interrupt service program, it does not need to interrupt the end of command EOI to 8259A. This way can only use only one 8259A in the system, and will not interrupt nested.

(2) Non automatic ending mode (EOI)

In this way, after the interrupt service has been completed, the ISR corresponding bit cannot automatically clear "0". The interrupt end command (EOI) must be sent to the 8259A before the end of the interrupt service program, and the corresponding bit is cleared "0", so it is called "non auto ending".

Non automatic termination is a common method, in which there are two command formats:

① The general way to remove the end of interrupt. ISR is not specified in the EOI command, but by the 8259A automatic cleaning the highest priority. In general, the first end of the interrupt service is the highest priority, this method is relatively simple, only for fully nested mode.

② Special interrupt ends. Specify the bits to be cleared in ISR in the EOI command. Used in a special full nested manner and in a loop priority manner.

CHAPTER 6 INTERRUPTION SYSTEM FOR MICROCOMPUTERS

In particular, in cascade mode, generally do not automatically interrupt the end, and need to use non automatic end interrupt mode, the cascaded from every interrupt handler at the end of two, must be sent over a EOI interrupt command, sent from a tablet, sent to the master.

5. Connection system bus mode

The 8259A connection system bus has two ways: buffering mode and non buffering mode

Buffer mode refers to the 8259A through the bus buffer and the system bus connected. In this way, connect the $\overline{SP/EN}$ pin of the 8259A to the enable terminal of the bus buffer. $\overline{SP/EN}$ signal output low start bus buffer.

The non-buffering mode is relative to the buffering mode where $\overline{SP/EN}$ is used as input. In single chip system, $\overline{SP/EN}$ is then connected to high; multi-chip 8259A, the master $\overline{SP/EN}$ is high, the slave $\overline{SP/EN}$ is low.

6. Interrupt inquiry mode

In the interrupt and query mode, the external device interrupt request signal to the 8259A interrupt request can be triggered, can also be a trigger level. But 8259A not through INT signal interrupt request signal to the CPU, because the CPU internal interrupt flag IF is "0", so the ban on CPU 8259A. CPU interrupt request to use the query to determine the software interrupt source, the peripheral interrupt service ability. Therefore, interrupt queries have interrupted the characteristics and features of queries.

6.3.5 Initialization of 8259A command word and initialization programming

8259A is a programmable interrupt controller, to carry out the program according to the service requirements and hardware connection before use. There are two groups of command registers inside the 8259A, a group is set in the 8259A initialization, call the initialization command word ICW, write another group in 8259A after initialization, called operation command word OCW.

1. 8259A initialization command word ICW

The 8259A has 4 initialization command registers. The $ICW_1 \sim ICW_4$. 8259A must write the initial command word before it starts to work, so that it works in a predetermined way.

The initialization command word formats are as follows:

(1) Initialization command word ICW_1

ICW_1 called chip control initialization command word. Write the address port of 8259A (Ie 8259A A0 must be 0). The command word format is shown in Figure 6-8, where each bit is defined as follows:

D_0 bit (IC_4): used to indicate whether or not to set ICW_4. $IC_4 = 1$, use ICW_4, $IC_4 = 0$, do not use ICW_4. and must use ICW_4 in 8086 systems, so the bit should be set to 1.

D_1 bit (SNGL): pointing out that there are 1 or more 8259A. SNGL = 1 in the system, there are only 1 8259A in the system. SNGL = 0, there are many 8259A cascaded in the system, the main bit and the D_1 bit of ICW_1 from the 8259A must be 0.

D_2 bit (ADI): does not work in 16 bit and 32 bit systems. It can be 0 or 1.

D_3 (LTIM): set the interrupt request signal in the form of. LTIM = 1, interrupt for the trigger level, high level. LTIM = 0, the interrupt request for edge trigger, rising. In level triggered mode, if the pulse before the arrival of $\overline{INTA}$, the IR input line high level does not hold, then the IRR register has set the IR bit is reset; ISR the corresponding bit will not set.

D_4: this bit is used as the identifier bit to distinguish the operation command word "OCW_2" and "OCW_3."

$D_5 \sim D_7$: does not work in 16 bit and 32 bit systems, can be 0 or 1. and is generally set to 0.

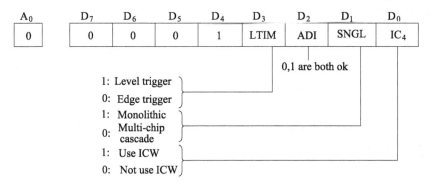

Figure 6-8　ICW1 format

(2) Initialization command word ICW_2

ICW_2 is the initialization command word that sets the interrupt type number and must be written to the 8259A's odd address port. The format is shown in Figure 6-9.

A_0	D_7	D_6	D_5	D_4	D_3	D_2	D_1	D_0
1	T_7	T_6	T_5	T_4	T_3	×	×	×

T7 ~ T3 indicate the upper 5 bits of the interrupt type

Figure 6-9　ICW_2 format

Only 5 (T7 ~ T5) are set up when the initialization of ICW_2 high, the lower 3 bits can be written on any value. An IR pin appears interrupt request signal, the interrupt type number is the high 5 bits of ICW_2, and the number of interrupt type low 3 bits introduced by the pins. For example: ICW_2 pins are initialized to 20H, then $IR_0 \sim IR_7$ interrupt type number corresponding to 20H ~ 27H, if the IR_3 pin appears request signal, the interrupt type is 23H. If ICW_2 is initialized to 25H, the interrupt type numbers are still 20H ~ 27H, IR_3 pin corresponding interrupt type number is also 23H.

(3) Initialization command word ICW_3

ICW_3 is command word, to write odd address port. When the system is 8259A cascade system, ICW_3 makes sense. Olny when ICW_1 SNGL bit is 0, ICW_3 is set up. In the cascade system, it generally one master 8259A, some slave of master are as 8259A. In response from the interrupt request process, $CAS_0 \sim CAS_2$ output, $CAS_0 \sim CAS_2$ if slave are as input. The master uses the 3 signal sends identification code to slave, in order to separate from addressing. Therefore, the master and slave ICW_3 settings are different. ICW_3 format is shown in Figure 6-10.

A_0	D_7	D_6	D_5	D_4	D_3	D_2	D_1	D_0
1	S_7	S_6	S_5	S_4	S_3	S_2	S_1	S_0

Si = 1, corresponding to IRi pin connected from slice

(a) main chip ICW_3

D_7	D_6	D_5	D_4	D_3	D_2	D_1	D_0
×	×	×	×	×	ID_2	ID_1	ID_0

ID2 ~ ID0 indicates the pin number of the slave chip

(b) Auxiliary chip ICW_3

Figure 6-10　ICW_3 format

In the master ICW_3, $S_7 \sim S_0$ corresponds to the connection on pins $IR_0 \sim IR_7$. If the IR_i pin has an slave, the corresponding S_i bit in the ICW_3 is "1"; the S_i bit is "0" if there is no connection.

The ICW_3 of slave, $D_7 \sim D_3$ does not use, can be set to any value. $D_2 \sim D_0$ value, called identification code (ID_2 to ID_0), indicating the signal from the chip connected to the main chip IR pin number, number 000 to 111, corresponding to the main chip $IR_0 \sim IR_7$ pin INT.

(4) Initialization command word ICW_4

ICW_4 in order to control the initialization word, write the odd address port. Only ICW_1 $D_0 = 1$, you need to write the ICW_4. After the ICW_3 (multi slice cascade system) or ICW_2 (monolithic 8259A mode). Otherwise, do not set the word.

ICW_4 specific format is as shown in Figure 6-11.

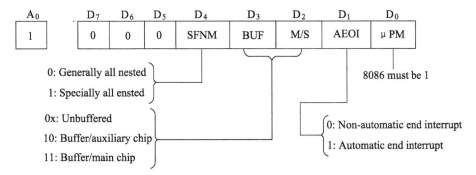

Figure 6-11 ICW_4 format

D_0 bit (μPM): μPM = 0, 8259A for 8 bit microcomputer; μPM = 1, 8259A for 16 bit microcomputer. In 8086 system, this bit must be 1.

D_1 (AEOI): set automatic end of interrupt. AEOI = 1, 8259A is in automatic end of interrupt. In this way, the second at the end of the pulse $\overline{INTA}$, the corresponding bit in current service register ISR······will automatically clear 0. So once interrupted, according to 8259A, the interrupt is over, allowing interrupt request at any level of nesting. AEOI = 0, 8259A is work in a non automatic end interrupt, and interrupt must end command in the interrupt service program.

D_2 bit ($M/\overline{S}$): master and slave chip identification bits. When D3 =1 (8259A works in buffered mode), this bit is significant. $M/\overline{S}$ =1, 8259A, is master. $M/\overline{S}$ =0, 8259A is slave.

D_3 (BUF): buffer mode enable bit. BUF =1, 8259A is in buffer hairstyle. Multi cascade system usually in this way, the 8259A buffer is being connected by bus and data bus, outputing enable signal. Then, D_2 bit in ICW_4 set master/slave BUF = 0, 8259A works in non buffer, 8259A and data directly connected. $\overline{SP/EN}$ is input, used as the master/slave control. At this time, ICW_4 D_2 has no significance.

D_4 bit (SFNM): special full nested bit. SFNM = 1, 8259A working in a special full nested manner, is used for main film in master/slave system. SFNM = 0, 8259A in fully nested mode, is used as slave for single or cascading systems from the chip.

$D_5 \sim D_7$: these 3 bits are always 0, as the ICW_4's identifying bits.

2. 8259A initialization process

The 8259A is a programmable chip that must be initialized to all 8259A in the system by initializing the command word before it works. Initialization is done in a certain order, and the initialization process is shown in Figure 6-12.

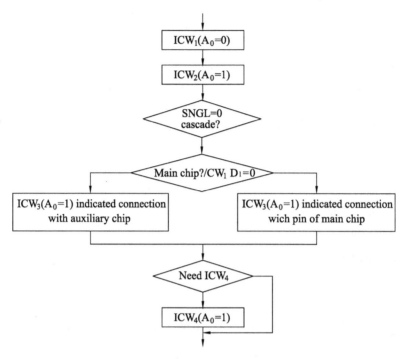

Figure 6-12 8259A initialization flowchart

As can be seen from the diagram, the initialization of 8259A must be according to the order of the $ICW_1 \sim ICW_4$. ICW_1 must be written to the even address port, $ICW_2 \sim ICW_4$ written to odd address port. For 8086, if only 1 pieces of 8259A, the initialization is not located in the ICW_3, setting the ICW_1, ICW_2 and ICW_4 according to the sequence. For cascade system, ICW_3 should be set both in master and slave, the format of the master-slave different.

3. Example of initialization of 8259A

[**Example 6-1**]   In IBM-PC, the 8259A port address were 20H and 21H, the work is as follows: single mode request signal using edge trigger mode, priority by completely nested, non automatic end interrupt, the system using the non buffer connection. Interrupt type is 08H~0FH.

Then the initialization sequence of 8259A is as follows:

```
     MOV, AL, 00010011B         ; ICW₁: edge triggered, monolithic systems
                                         requiring ICW₄
OUT 20H,AL
MOV, AL, 00001000B              ; ICW₂: type number starts from 08H
OUT 21H,AL
MOV, AL, 00000001B              ; ICW₄: completely nested, non buffered, non
```

CHAPTER 6   INTERRUPTION SYSTEM FOR MICROCOMPUTERS

```
                          auto ended
OUT 21H,AL
```

[**Example 6-2**]   IBM PC/AT machine using two 8259A management interrupt, the hardware connection shown in Figure 6-13. $\overline{SP/EN}$ connected to the high level is master, $\overline{SP/EN}$ then slave connected to low level. Master 8259A in $A_9 A_8 A_7 A_6 A_5 = 00001$, chip select signal valid, I/O address between 20H ~ 3FH are selected this piece of 8259A, ROM BIOS using address 20H and 21H. Slave 8259A chip select address is 0A0H ~ 0BFH, commonly used 0A0H ~ 0BFH. IR_0 of the main chip 8259A is used for the daily clock interrupt of the microcomputer system, IR_1 connected to the keyboard interrupt request signal, IR_2 connected to the slave. IR_0 in slave 8259A with real clock, IR_6 connected hard disk interrupt.

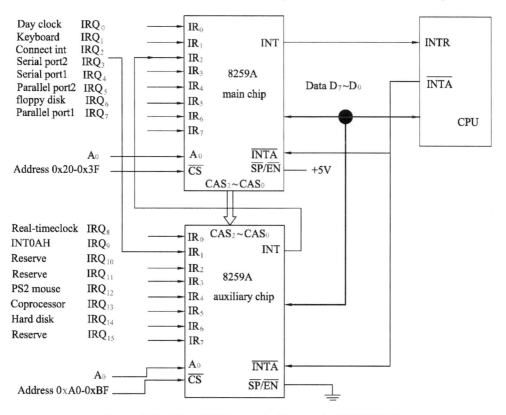

Figure 6-13   Two 8259A cascade diagram in IBM PC/AT

The initialization procedure of the master 8259A is as follows:

```
MOV, AL, 00010001B    ; ICW₁: edge triggered, multi slice cascade,
                        requiring ICW₄
OUT 20H,AL
MOV, AL, 00001000B    ; ICW₂: type number starts from 08H
OUT 21H,AL
MOV, AL, 00000100B    ; the ICW₃:IR₂ pin is connected from the chip
OUT 21H,AL
MOV, AL, 00010001B    ; ICW₄: special nested mode, non buffering, non
                        automatic ending
```

```
OUT 21H,AL
```

From the initialization section of the slice 8259A, the program runs as follows:

```
MOV, AL, 00010001B      ; ICW₁: edge triggered, multi slice cascade,
                                requiring ICW₄
OUT 0A0H,AL
MOV, AL, 01110000B      ; ICW₂: type number starts from 70H
OUT 0A1H,AL
MOV, AL, 00000010B      ; ICW₃: the IR₂ pin connected from the chip to
                                the main slice, and the ID code is 02H
OUT 0A1H,AL
MOV, AL, 00000001B      ; ICW₄: completely nested, non buffered, non
                                auto ended
OUT 0A1H,AL
```

6.3.6 Operation command word and application of 8259A

8259A has three operation command word $OCW_1 \sim OCW_3$. Command word is in 8259A after the initialization settings in the application program. No set order, but strict rules on the port address: OCW_1 must be written to the odd address port, OCW_2 and OCW_3 written to the address port.

1. Operation command word "OCW_1"

(1) The format and meaning of OCW_1

OCW_1 is called interrupt mask operation command word, shielding operation of 8259A interrupt request signal, written to the odd address port. The specific format is shown in Figure 6-14.

A_0	D_7	D_6	D_5	D_4	D_3	D_2	D_1	D_0
1	M_7	M_6	M_5	M_4	M_3	M_2	M_1	M_0

$M_i = 1$, the interrupt corresponding to the IR_i pin is masked
$M_i = 0$, open the corresponding IR_i pin interrupt request

Figure 6-14 OCW1 format

If an pin IR of the 8259A acts as an external interrupt request signal, the corresponding bits of the OCW_1 must be cleared 0, in the application

(2) Application of OCW_1

A system of 3 interrupt sources are respectively connected to the 8259A IR_0, IR_2 and IR_3. In order to make the 3 interrupt request can apply to the CPU through the 8259A interrupt, after initialization, the 8259A must open the corresponding interrupt mask bit. If even address of 8259A in system is 80H, the odd address is 81H. Interrupt word is written by odd address, procedures are as follows:

```
MOV, AL, 11110010B      ; 0 bit allowed interrupt; 1 bit interrupt is
                          blocked
OUT 81H,AL
```

If the system adds a IR_1 interrupt request, in order not to affect other interrupt source, we should first read the original interrupt mask, and then use the AND command to clear the mask corresponding to the IR_1 0 and then write it to the OCW_1. The code follows:

```
IN AL,81H
AND, AL, 11111101B      ; only D₁ bits in OCW₁ are cleared "0""
OUT 81H,AL
```

When all the interrupts in IR_1 are complete, the external request should be shielded from the D_1 location of OCW_1. The code reads as follows:

```
IN AL,81H
OR AL, 00000010B        ; only the D₁ position "1" in OCW₁ does not affect
                          other interrupt sources
OUT 81H,AL
```

2. Operation command word "OCW_2"

(1) The format and meaning of OCW_2

OCW_2 is used to set the priority cycle mode and interrupt the end of the operation of the command word, written to even address port. The specific format is shown in Figure 6-15.

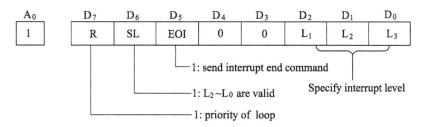

Figure 6-15 OCW_2 format

D_7 bit (R): the priority loop controller. R = 1 is the cyclic priority, and R = 0 is the fixed priority.

D_6 bits (SL): $L_2 \sim L_0$, enable bits. SL = 1, $L_2 \sim L_0$ valid; SL = 0, $L_2 \sim L_0$ invalid.

D_5 bit (EOI): end of interrupt command bit. EOI = 1, in the end of non automatic interrupt mode, when the interrupt service program is finished, the interrupt is cleared 0. In the corresponding position of the ISR register by making the EOI set 1 to notify the 8259A. EOI = 0, in interrupt auto ending mode, do not need to send end of interrupt command.

$D_4 D_3 = 00$, characteristic bits.

$D_2 \sim D_0$ bits ($L_2 \sim L_0$): specify the interrupt level, 000 ~ 111, corresponding to 0 ~ 7 interrupt.

OCW_2 mode word function includes two aspects: on the one hand, through the R bit can determine whether the 8259A priority cycle, through the combination of R, SL and $L_2 \sim L_0$ determine the current minimum priority; on the other hand through the EOI bit General interrupt end command, through EOI, SL and $L_2 \sim L_0$ combination to achieve a special interrupt end of the command. Table 6-2 summarizes the functions of OCW_2. The most commonly used is the general interrupt termination command, which clears the ISR bit with the highest priority in the interrupt service register.

Table 6-2　Function of OCW$_2$

R	SL	EOI	00	L2	L1	L0	Function
1	0	0	00	0	0	0	Priority auto-loop mode
1	1	0	00	L2	L1	L0	Priority special loop mode, L2L1L0 is the lowest priority
1	0	1	00	0	0	0	The interrupt ends the command and still uses the priority loop mode
1	1	1	00	L2	L1	L0	The interrupt ends the command and still uses the priority special loop mode
0	1	1	00	L2	L1	L0	Set the special interrupt end command to clear the L2L1L0 corresponding bit
0	0	1	00	0	0	0	Issued a general interrupt end command

(2) OCW$_2$ application

End of interrupt (EOI) command is a command with most use of OCW$_2$. When the interrupt is set to a non automatic end, send interrupt end command before the interrupt service program ends. If 8259A even address in the system is 80H, the odd address is 81H. EOI command as follows:

```
MOV AL, 00100000B      ; EOI, 20H, OCW₂'s EOI position 1, that is, the
                         generic command
OUT, 80H, AL           ; output from a bipartite address
```

If it is 8259A cascade system, before the end of slave interrupt service program two EOI commands must be sent another to master. Assuming the master address is 20H, 21H, slave address is 0A0H, 0A1H. Procedures are as follows:

```
MOV AL, 00100000B
OUT, 0A0H, AL          ; start from the slice EOI command
MOV AL, 00100000B
OUT, 20H, AL           ; Post Master EOI command
```

3. Operation command word "OCW$_3$"

(1) The format and meaning of OCW$_3$

Operation command word OCW$_3$ has three functions: ① set and revoke the special shielding; ② set the interrupt query; ③ set the 8259A internal register read command. OCW$_3$ must be written to the even address port, the specific format shown in Figure 6-16.

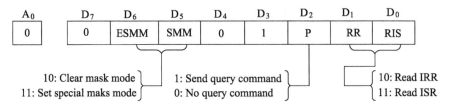

Figure 6-16　OCW3 format

D$_7$ bit is always 0.

D$_6$ bit (ESMM): special mask mode enable bit. ESMM = 1, SMM bit valid; ESMM = 0,

SMM bit invalid.

D_5 (SMM): a special shield mode. SMM = 1, set the 8259A in special shield mode; the system back to the original work. In the interrupt service process, interrupt request may be required to open of lower priority than current interrupt request, but in fully nested mode, low priority interrupt will be forbidden. Set up a special shielding method can solve this problem. The setting method is: the first set OCW_3 ESMM = SMM = 1, then OCW_1 will be set up to mask current service interruption, as long as the CPU IF = 1, you can open the service in addition to being Interrupt requests (including low priority requests) outside the interrupt.

$D_4 D_3 = 01$, which is OCW_3's identity bits.

D_2 bit (P): query mode bit. P = 1 denotes 8259A work in the priority query mode. In this mode, the CPU does not rely on the receive interrupt request signal to enter the interrupt processing process, but by issuing the query command word to read the query word to obtain the peripheral interrupt request information. Before CPU each read the query word, it must first output OCW_3 (P = 1) query command to 8259A even address, and then content from the even address port is the query word. The format of the query word is shown in Figure 6-17.

A_0	D_7	D_6	D_5	D_4	D_3	D_2	D_1	D_0
0	I	–	–	–	–	W_2	W_1	W_0

Figure 6-17 I = 1 means interrupt request $W_2 \sim W_0$ highest priority interrupt source code

I = 1, said the interrupt request, when $W_2 W_1 W_0$ is pending interrupt interrupt source in some of the highest priority interrupt source encoding. For example, if the query word is 10000010B, it indicates that the current highest level of the request is IR_2, and you can go to the IR_2 service program to execute. I = 0 indicates that there is no interrupt request at present.

D_1 bit (RR): read the internal register enable bit. RR = 1, you can read the contents of the internal 8259A ISR or RR. IRR = 0, you can't read the internal registers.

D_0 bit (RIS): the internal register select bit. RIS = 1, reads the contents of the ISR register. RIS = 0, reads the contents of the IRR register.

(2) Application of OCW_3

The main function of OCW_3 is to set up special shielding and query words. If the 8259A even address of a system is "80H", the "odd address" is 81H. the main applications of OCW_3 are as follows:

① Interrupt inquiry mode

```
    MOV, AL, 00001100B; P bit =1 of OCW₃        ; set query mode
    OUT 20H,AL
    IN AL,20H
;The read content is an interrupt query word. If the highest bit is 1, the
interrupt request is indicated, and the highest priority number is represented by
a low 3 bit.
```

② Query the status of the ISR register

```
    MOV, AL, 00001011B        ; OCW₃'s RR bits and RIS bits are 1; query the ISR
                                registers
```

```
    OUT 20H,AL
    IN, AL, 20H              ; the read content is the state of the ISR
                               register
```

③ Querying the status of the IRR register

```
    MOV, AL, 00001010B       ; the RR bit of OCW₃ is 1, the RIS bit is 0, and the
                               query IRR register
    OUT 20H,AL
    IN, AL, 20H              ; the read content is the state of the IRR
                               register
```

6.4 Interrupt application example

We have a good understanding of the working principle and the working process of the interrupt system. In order to let readers know the design process of interrupt transmission, an example of interrupt application is given now.

[**Example 6-3**] Assuming a certain application system control 8 LED through a set of 8 switch state, corresponding LED lights when a switch is closed. The requirements of the 8 switch state all the setup is complete, can light up, when the switch is 0FFH, end the program. Complete hardware and software design.

1. Hardware circuit design

The hardware circuit design determines the data transmission mode according to the demand, determin the main function of the interface and the components. In this case, switch and a LED is the most basic input and output devices, the switch can be directly through the three state buffers of the 74LS244 and CPU data lines, LED by the latch and 74LS273 connected bus. Can unconditional transmission that spoke in accordance with chapter sixth achieve this function? According to the unconditional transfer principle, state of the switch can be read at any time according to the value of LED, a switch state change will immediately react to LED. While the system requirements the setup of 8 switch state is complete, LED can light up the corresponding switch, not intermediate state from one state to another state transformation led to change. For example, to the value of the switch from the 00000001B set to 00001111B by 00000011B, 00000111B two the intermediate state, the two state not reflect to LED. So the unconditional transmission mode to achieve the function is by not to query or interrupt transmission. In this case the interrupt mode for this function.

According to the requirements, when the user sets the switch data, it can be generated by a single pulse generator to send a positive pulse signal to the 8259A interrupts request; 8259A sent the request signal to the sin INTR of CPU after the received pin interrupts request signal; The CPU issues an INTA acknowledge signal under permission to read to internet, retrieves the interrupt class number from 8259A, and proceeds to the corresponding interrupt service routine. As shown in Figure 6-18, the interface circuit includes data input port 74LS244, data output port 74LS273, 8259A control circuit and address decoding circuit with. 8259A data line and 8086CPU $D_0 \sim D_7$ connected, CPU A_0 must be 0,8259A in order to exchange data with the 8086CPU. So the address decoder 80H output signal must be 8086CPU A_0 phase or after

CHAPTER 6 INTERRUPTION SYSTEM FOR MICROCOMPUTERS

8259A with the $\overline{CS}$ connected, 8259A A0 signal connected with the 8086CPU A1. It can be seen, 8259A two port addresses were 80H and 82H. The data input port address is 0A0H and the data output port address is 0C0H.

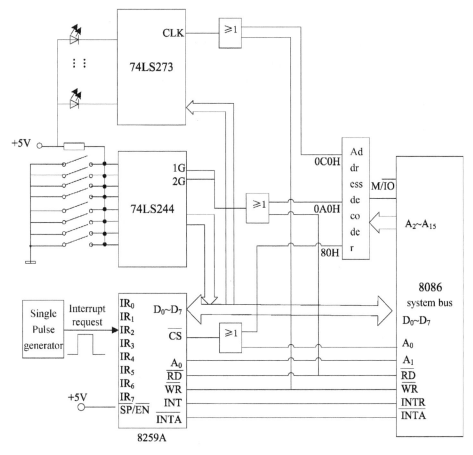

Figure 6-18　Interrupt mode input and output interface schematic

2. Software programming

Software programming includes two modules: the main program module is to the main completion of a variety of initialization work, interrupt service subroutine module to complete the data input and output.

(1) Main program

In order to enable interrupt service program to execute correctly, the main program should do the initialization of the interrupt system. The initialization of the main program usually includes the following 5 parts:

① CPU initialization. It includes initialization of the system segment register, initialization of the stack, and initialization of the interrupt vector table.

② Interrupt controller 8259A initialization. It sets the mode of operation of 8259A, set the interrupt type number, set priority management, set the end of the interrupt, clear the 8259A mask bits and so on.

③ Initialization of the external device interface. The interface is restored to its original state. For programmable interfaces, initialization, set of work modes, open interrupts, shielding words, etc. are set up.

④ The initialization of the interrupt service routine. It sets the buffer pointer, counter, status bit, etc. used by the interrupt service routine.

⑤ Open CPU interrupts. It sets the IF flag to 1 by means of the STI command.

Note: due to the randomness of external interrupt, the interrupt service program pointer, counter, flag etc. can only be placed in the memory unit. Enter the interrupt service program, the protection of relevant registers, the pointer, counter, sign can be loaded in registers for use. If values loaded into the register changed during the interrupt service program, before the end of interrupt service program, it need to be stored in the corresponding memory unit.

For the output process, the first output should be started in the main program, otherwise the interrupt output request signal will not be generated.

In the main program to complete the interrupt initialization, it can perform other tasks of the system program. After the input of interruption, the main program do the work according to the needs analysis. If the end of processing input data, save data to the hard disk, and the corresponding position to close the 8259A shielding interruption.

(2) Interrupt service routine

The interrupt service routine usually consists of the following steps:

① Protect the scene. Press all registers that are used or may change values in the interrupt service routine to stack protection.

Note: the pointer, buffer, etc. used by the interrupt service program are stored in the memory cell. In order to load the pointer and access data, the segment register needs to be reloaded. Therefore, the protection site should include the segment register.

② Open interrupt. By using the STI command, the IF flag is set again to 1, allowing for higher priority or more urgent interrupts.

③ Load data buffer pointer, counter, in the interrupt mode, their values should be stored in memory at ordinary times, before using to load the corresponding register, in order to facilitate data access.

④ Input and output processing. For the input process, the data read from the input port, check the correctness of the data (such as parity), data will be stored in the buffer. The pointer and the value of the counter and write memory, check whether the input end, if end, set the corresponding symbol. For the output process, it is to the output port next to the output data, and modify the corresponding pointer and counter.

⑤ Off interrupt. Set the IF flag to 0 by means of the CLI command to close the interrupt to avoid unnecessary interrupt nesting.

⑥ Send the EOI command to 8259A. Clear the corresponding bits of the ISR register in the 8259A in response to a sibling or low-level interrupt request.

⑦ Restore the scene. Recover the contents of each register according to the principle of "advanced out".

⑧ Break return. Returns to the interrupted program with the IRET command.

(3) Complete source code

According to the hardware circuit design of 8259A port address for 80H and 82H, three

CHAPTER 6 INTERRUPTION SYSTEM FOR MICROCOMPUTERS

state buffers of the 74LS244 address for the 0A0H latch 74LS273 port address for the 0C0H. 8259A initialization for a monolithic, edge trigger, fully nested, non automatic end interrupt, interrupt type number is 40H ~ 47H, the IR_2 pin as an input pin interrupt request therefore, the interrupt source type is 42H. Main program code is as follows:

```
;Data segment definition
DONE DW                  ; interrupt completion flag
;Code segment
;Interrupt vector table initialization
MOV AX,0
MOV ES, AX               ; the interrupt vector table segment base address is
                           0000H
MOV, BX, 42H * 4         ; preset interrupt vector address in BX; interrupt
                           type number is 42H
MOV AX,OFFSET INTRPROC
 MOV, ES:[BX], AX        ; the address of the vector table is low and the
                           offset address is stored
MOV AX,SEG INTRPROC
 MOV, ES:[BX + 2], AX    ; base address of high address storage in vector
                           table
8259A initialization
MOV, AL, 00010011B       ; ICW₁: edge triggered, monolithic systems requiring
                           ICW₄
OUT 80H,AL
MOV, AL, 01000000B, ICW2 : type number starts from 40H
OUT 82H,AL
 MOV, AL, 00000001B      ; ICW₄: completely nested, non buffered, non auto
                           ended
OUT 82H,AL
MOV, AL, 11111011B       ; open IR₂ request interrupt mask bits
OUT 82H,AL
Interrupt service routine initialization
MOV, DONE, 0             ; completion flag 0,
STI
WAIT:, CMP, DONE, 1      ; DONE =1 means all interrupts completed
JNZ WAIT
MOV, AX, 4C00H           ; returns DOS
INT 21H
Interrupt service subroutine
INTPROC PROC FAR
PUSH AX                  ; protect the scene
PUSH DS
STI                      ; open interrupts; allow interrupt nesting
MOV AX,SEG DONE
MOV DS,AX
IN, AL, 0A0H             ; read a data from the input port
OUT, 0C0H, AL            ; output the data from the output port and light the
```

```
                        LED
    CMP, AL, 0FFH       ; determine whether the switch to read is 0FFH
    JNZ EXIT
    MOV DONE, 1         ; the input data is 0FFH, and the end flag is set
    IN, AL, 82H         ; set the interrupt mask bit of IR2
    OR AL,00000100B
    OUT 82H,AL
    EXIT: CLI           ; off interrupt, ready to return
    MOV, AL, 20H        ; send the EOI command
    OUT 80H,AL
    POP DS              ; restore the scene
    POP AX
    IRET                ; break back
    INTPROC ENDP
```

Exercise 6

6.1 What are interrupts? What are the two categories that interrupt sources can be divided into?

6.2 A complete interruption process is briefly described.

6.3 When there are multiple interrupts sending interrupts to 8086, in what order does it respond to interrupts?

6.4 What is an interrupt vector? What is an interrupt vector table? Assuming that the interrupt type number is 43H and the interrupt vector is 3210H:4567H. Please draw a diagram to illustrate how to store the interrupt vector.

6.5 Suppose that interrupt type 9 interrupts the service program's first address to INTR_9, and writes the program segment that sets the interrupt vector table.

6.6 What is the function of the 8259A interrupt request register IRR, the interrupt mask register IMR and the interrupt service register ISR?

6.7 If 8259A IR_1, IR_4 and IR_7 are connected with three interrupt sources, IR_4 and IR_7 also issued the interrupt request signal, in the interrupt service period, IR_1 also issued an interrupt request. If 8259A works in full nested mode, port address for the 20H and 21H.

(1) Code open 3 interrupt source mask bit CPU.

(2) In what order response interrupt three interrupt sources?

6.8 If the interrupt type number is set from 20H ~ 27H when the 8259A initializes, what is the corresponding interrupt type number for IR_3?

CHAPTER 7
PROGRAMMABLE INTERFACE CHIP

【Abstract】 This chapter mainly introduces the programmable parallel interface chip 8255A, programmable serial interface chip 8251A and timer/counter chip 8254 internal structure, working principle and programming applications.

【Learning Goal】
- Master the internal structure, working principle, interface design method and programming of parallel interface chip 8255A.
- Master the classification of serial communication, synchronous communication and asynchronous communication.
- Familiar with RS-232C serial communication protocol.
- Master the design method of 8251A interface and the preparation of control program.
- Master programmable timer/counter chip 8254 structure, 6 modes of work and related programming applications.

7.1 Programmable parallel interface chip 8255A

The data transmission between CPU and peripherals are data transmission . The data transmission between CPU and interface is parallel, which is usually to word as a unit, a transmission number, usually 8, 16 or 32; and the data transmission interface and the peripherals can be divided into two types: serial transmission and parallel transmission. The serial transmission is one of a data transmission in a transmission line, and a plurality of parallel transmission data in the transmission line which is usually in a word is a number of units for simultaneous transmission. Compared with the serial transmission, parallel transmission needs more transmission lines, but the speed of transmission is higher, especially in high speed and short distance occasions.

To achieve the parallel transmission interface called parallel interface, parallel interface consists of a programmable parallel interface and programmable parallel interface. Programmable parallel interface is usually composed of three state buffers and data latch structure, control of the interface is relatively simple, but to change its function must change the hardware circuit. The interface has been discussed in Chapter sixth. The biggest characteristic is its functional programming interface can be programmed to set and change, so it has great flexibility.

8255A is a parallel interface chip produced by Intel programmable, 40 pins, 3 8-bit ports,

3 kinds of operating modes. Each port function can be selected by the software, the use of flexible, versatile. 8255A can be used as intermediate host and various peripheral interface circuit connection.

7.1.1 The internal structure of 8255A

Bridge 8255A as host and peripherals connected, on the one hand to be connected with the host, on the other hand to provide a variety of peripherals need. So from the functional perspective, the signal must have interface connected with the CPU interface and peripherals. At the same time it has the programmable characteristic, its structure and function can be realized by software programming, the interior must with the logic control part. The internal structure of 8255A is divided into 3 parts: the connection part, the CPU part, the control part and peripherals. As shown in Figure 7-1.

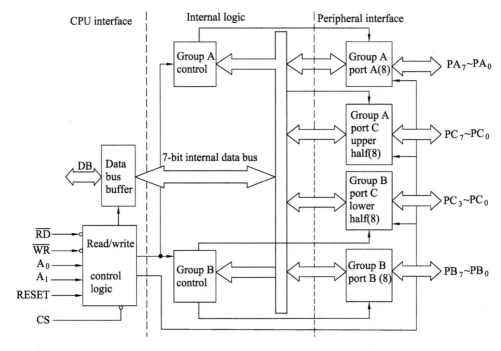

Figure 7-1　Internal structure of 8255A

1. Connect to the CPU section

The part of the connection between the 8255A and the CPU consists of two modules, one is the data bus buffer and the other is the read/write control logic.

(1) Data bus buffer. 8-bit bidirectional data bus three state buffer CPU and 8255A is the only way which must be passed for the exchange of information, also is essential function of all interface chip connected to CPU. Before the start of 8259A data transmission, CPU through the data bus buffer 8255A initialization control word is sent to 8255A, and then through the data bus buffer for input and output data.

(2) Read and write control logic. All of CPU operations on 8255A are in the control of read/write control logic circuit to achieve the control command. The circuit receives control command from the CPU, and sends operation command according to the command to chip the

function parts.

2. Peripheral interface part

8255A has three 8-bit ports and peripherals connected, respectively, port A, port B and port C. which serves as the input buffer is connected with the input device, or as the output latch and output devices connected. The port A and port B as the main data input/output. Port C in addition to the 8-bit port A data input/output port, also can be divided into two 4-bit ports, and for port A and port B cooperation, works as the state port or the control port.

3. Internal control logic

8255A internal control logic is divided into two groups: A group and B group, control three ports (A port, B port and C port) working group. A control logic control port A and port C high 4 bits; group B control logic control port B and port C low 4 bits.

7.1.2 Pin function of 8255A

The 8255A is a dual in line package with 40 pins, and the pin signal, shown in Figure 7-2, can be divided into pins connected to the CPU and pins connected to the peripherals.

1. Pins connected to the CPU

$\overline{CS}$: The chip select pin, input, active low. When $\overline{CS}=0$, the chip is selected, allowing 8255A to communicate with CPU; when $\overline{CS}=1$, 8255A can't do data transmission with the CPU.

$\overline{RD}$: Read the control pin, output, active low, when $\overline{RD}=0$ and $\overline{CS}=0$, CPU reads the status or data from the 8255A.

$\overline{WR}$: Write the control pin, input, active low, when $\overline{WR}=0$ and $\overline{CS}=0$, CPU writes data or controls words to 8255A.

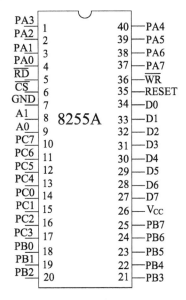

Figure 7-2　8255A pinout

$D_0 \sim D_7$: 8-bit, three state bidirectional data pins, which are connected with the system data bus, realize data, control the transmission of word and status information.

A_0, A_1: The address input pin is used to select the four ports within the 8255A.

RESET: Reset input pin. When high level is active, all internal registers (including control registers) are cleared, and all I/O ports are set to input mode.

Usually, $D_0 \sim D_7$ and $\overline{RD}, \overline{WR}$ of 8255A directly with the system data bus and a read and write control signals connected to system usually consists of high address signal decoding. A_0, A_1 system and low address lines.

If the system is used in 8086 CPU, the data bus line for 16. CPU in the transmission of data, always low 8-bit of data sent to my address port, high 8-bit of data sent to the odd address port. When the 16-bit data bus memory or I/O interface chip only has a 8-bit data bus and 8086 linked, it can be connected to the 8-bit data bus, can also be connected to the lower 8-bit of the data bus. In the actual design of the system, for the sake of convenience, often the data line of these chips $D_7 \sim D_0$ to the system data bus low 8, so that each port 8086 in the system should be for my address, 8 and A0 interface chip 086 address lines, select the chip port with other low address line. Only 8088 and 8 data lines, there is no address parity problem, 8088 A0, A_1 can be used to select the interface chip inside the port.

Table 7-1 lists the read and write operations control for each port.

Table 7-1 Port read write operation and pin signal relation table

$\overline{CS}$	$\overline{RD}$	$\overline{WR}$	A_1	A_0	Port selection and operation
0	1	0	0	0	Writes data to port A
0	1	0	0	1	Writes data to port B
0	1	0	1	0	Writes data to port C
0	1	0	1	1	Writes a control word to the control port
0	0	1	0	0	Read data from port A
0	0	1	0	1	Read data from port B
0	0	1	1	0	Read data from port C
0	0	1	1	1	no-operation
1	×	×	×	×	prohibition of use
0	1	1	×	×	no-operation

2. A pin that is connected to a peripheral

$PA_0 \sim PA_7$: The 8 pin signal of port A.

$PB_0 \sim PB_7$: The 8 pin signal of port B.

$PC_0 \sim PC_7$: The 8 pin signal of port C.

3. Power and ground wire

Vcc, GND: Power and ground. Vcc generally take +5V.

7.1.3 How does 8255A work

1. Mode 0-Basic I/O mode

The 0 is a basic I/O mode, in this way, the three ports can be programmed for input or output, not for the fixed contact signal response. Its basic functions can be summarized as follows:

(1) May have two 8-bit ports (port A and port B) and two 4-bit ports (the upper half of port C port and the lower half of port C).

(2) Each port can be set as input or output, and there are 16 kinds of I/O combinations.

(3) The output has latch capability, the input has only buffer capacity, and no latch function.

Ports working in mode 0 is applied to unconditional transmission and query mode. As the unconditional transfer mode, CPU can use the I/O instructions and peripherals data exchange. For example, a 8255A port in 0 as a keyboard or control LED digital tube interface, data input and output to no way, do not need to contact the signal interface circuit. As the query mode, need to use the control/status signal as some C ports. For example, 8255A A or B work in 0 way as the printer interface, need to use a certain shape as control/C port state signal.

2. Mode 1-strobe I/O mode

The 1 is a gated I/O mode. Only the A port, B port can work in this way, regardless of input or output to fixed requisition of some pins in port C as a contact signal, signal in port C not occupied for input or output. Data I/O in mode 1 have the ability of latching. The port also provides an interrupt request logic and interrupt enable flip-flop work in 1 way. The port for transmitting inquiry mode and interrupt mode.

(1) Mode 1 input

A port and B port work, respectively, in mode 1 input, respectively, C mouth 3 fixed pin as liaison signal, as shown in Figure 7-3, the three signals are defined as follows:

$\overline{STB}$: the strobe input is active low. The input signal sent by the peripheral device latches the data sent by the input device to the 8255A input latch when it is valid.

IBF: the input buffer is full, and the high level is the contact signal for the 8255A output. When it is valid, it indicates that the data is locked in the input latch. It is set by the low level of the $\overline{STB}$ signal and reset by the rising edge of the $\overline{RD}$ signal.

INTR: the interrupt request signal, high level, the output signal of 8255A, an interrupt request to CPU, CPU reads the data provided by the peripheral. 8255A interrupt request condition: INTE =1 (interrupt enable), $\overline{STB}$ inputs a complete negative pulse into high level and IBF =1. Interrupt request signal from the falling edge of $\overline{RD}$ reset.

INTE: interrupt enable bit. When INTE =1, the interrupt, 8255A can be sent to the CPU interrupt request signal. INTE =0, interrupt for bidden. The control bit does not correspond with the special external pin, group A and group B respectively controlled by PC_4 and PC_2 set/reset.

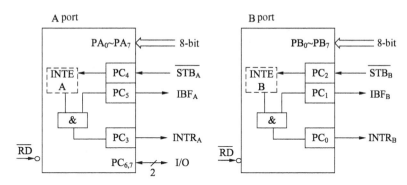

Figure 7-3 8255A mode 1 enter the contact signal

In order to better understand the working process of the 1 input, we cite to illustrate a popular example. Assume information transmission in undercover officers and liaison officers during the war, is often doesn't met at the agreed place to sign messages between them. Now, we assume that the contact is a member of CPU (an undercover intelligence). Personnel is used for input peripherals (to collect Intelligence Liaison), 8255A chip is connected to the CPU and agreed room peripherals. Because the two sides can not meet, pre arranged to see place to do an agreement, such as whether the house lights. Once at the agreed provisions undercover intelligence room, open the switch, the room lights; once the liaison to the intelligence, close the switch, the lamp, a transfer can be carried out after the end of the intelligence, and so on. Then the switch in the room is just like our 8255A chip $\overline{STB}$, peripheral gating, and IBF is the lights in the room, controlled by a switch, the signal is CPU to see, once meet again after reading the information, read and reset the IBF signal, a data read end.

Take A port as an example to introduce the process of sending a data to CPU by 8255A through 1:

① After the peripheral sends data to the data line $PA_0 \sim PA_7$ of the A port, the strobe signal $\overline{STB}_A(PC_4)$ is valid, and the data enters the input buffer of the A port.

② $IBF_A(PC_5)$ of the A port is valid, informing the peripheral and the CPU that the A port receives a valid data.

③ $INTR_A(PC_3)$ of the A port is valid, informing the CPU of the data in the A port by interrupting it.

④ CPU reads A, and data enters CPU.

⑤ IBF_A and $INTR_A$ become invalid and notify peripheral to send the next data.

(2) Mode 1 output

A port and B port work, in the way 1 output, respectively, C mouth 3 fixed pin as a contact signal, as shown in Figure 7-4, the three signals are defined as follows:

$\overline{OBF}$: the output buffer full, active low, 8255A is output to the contact signal peripherals. Said CPU has been sent to the designated port of the output data, the data can be removed outside. It is a rising signal along the $\overline{WR}$ set to "0" (effective), by the falling edge of the $\overline{ACK}$ signal is set to "1" (invalid).

$\overline{ACK}$: the peripheral response signal is active low. Indicates that the data CPU output to the 8255A has been removed by the peripheral.

INTR: the interrupt request signal, high effective. The data has been removed to CPU peripherals, request the interrupt request output data. The condition is: INTE (interrupt) = 1, $\overline{ACK}$ input a complete negative pulse into high level and $\overline{OBF}$ also changed from low to high. The interrupt request signal by down reset.

INTE: the interrupt enable bit is the same as the mode 1 input. Group A and group B are controlled by PC_6 和 PC_2 bit setting/reset respectively.

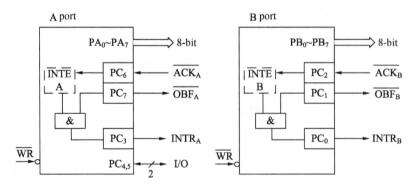

Figure 7-4 8255A mode 1 output contact signal

The following also take A port as an example, introduced CPU through 8255A way 1 send a data to peripheral process:

① The CPU executes the OUT instruction and writes the data to the output buffer of the A port.

② When valid data is entered on the port A data line $PA_0 \sim PA_7$, the $\overline{OBF}(PC_7)$ is valid, informing the peripheral CPU that an efficient data has been input to the A port, and that the peripheral can fetch data from the port A.

③ When the peripheral takes the data, it sends $\overline{ACK}(PC_6)$ signals to the 8255A and tells the A that the peripheral has taken the data.

④ Port A $\overline{OBF}(PC_7)$ is invalid. This indicates that the port A data has been removed by the peripheral.

⑤ $INTR(PC_3)$ valid to notify the CPU interrupt, and then output data to the port A.

3. Mode 2-bidirectional transmission mode

Mode 2 combines the gating, input, and output functions of mode 1 into a bidirectional data port to transmit data and receive data, for example, for bidirectional parallel communication between two processors.

Only the 8255A port A can work in 2 ways, as shown in Figure 7-5, 5 pin fixed port C signal acquisition, the meaning and the way of 1, is actually the A port in 1 combinations of input and output. Therefore, the same applies to the query transfer mode and interrupt mode.

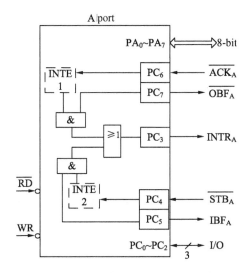

Figure 7-5 8255A mode 2 contact signal

Port A works in 2 ways, the input and output sharing an interrupt request signal. The input and output interrupt trigger $INTE_1$ and $INTE_2$ respectively by PC6 and PC4 set/reset to control. Therefore, when using the interrupt mode to transmit data, if the input and output while allowing the interrupt in the interrupt service interruption. The program reads the C port state for the detection of IBFA and $\overline{OBF_A}$ of the state, to further determine input or output interrupt interrupt.

When port A operates at mode 2 (C high 5 bits), B port can operate in the mode 0 (C port is low, 3 bits can work in mode 0), or mode 1 (C port low 3-bit is requisitioned), either as an input port or as an output port.

7.1.4 Control word for 8255A

8255A the control word two: control word and port C according to position/reset control word (also called a control word). The two control words are written to the control port. 8255A through D_7 identifies the write control port which in the end is a control word. Mode control word D_7 is always 1, and the C port according to position/reset the control word of D_7 is 0, so, D_7 is known as a logo.

8255A is a general-purpose parallel interface chip, in the specific application, it should be selected according to the actual situation. So the work use must first initialize the write control word to specify their way of working. If you need to interrupt, but also with C according to the position/reset control word will allow the 1 interrupt trigger INTE or clear 0. Initialization is complete, you can on the 3 data port for reading/writing.

1. Mode control word

Control word for work mode and direction of input and output 3 data set the port of 8255A. Three ports are divided into two groups: port A and port C high 4 bits as group A, port B and port C low 4 bits as the group B. D_7 is feature bit, which is always 1. $D_6 \sim D_3$ for group A set $D_2 \sim D_0$ to set group B. Select the control word as shown in Figure 7-6.

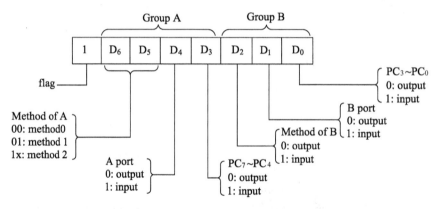

Figure 7-6　8255A mode select control word

In mode 1 and mode 2, port C bits that are fixed for contact with the signal acquisition, port A or port B. The word does not provide C way of working, because the C was not the only export requisition work in 0 ways.

The initialization of the 8255A first is to set the mode of operation and the input/output mode of the 8255A, that is, to write the mode control word to the control port through the output command.

[**Example 7-1**]　Assume that the use of 8255A port A output in mode, port B in mode 0 input, C port high 4 bit input, C port low 4 bit output. Suppose 8255A's 4 port address is 60H, 61H, 62H, 63H (the following example port address the same).

Then the control word is: 10001010B approach 8AH.

Initialize the program as:

```
MOV AL,8AH
OUT 63H,AL
```

If the 8255A control port is written to the 10100111B, the 8255A three ports work as follows: port A output in mode 1, port C half output, port B input in mode 1, port C half of the input.

2. Port C by position/reset control word

When the port C is set to output mode, it is often used for control purposes to send control signals. For example, a lamp is controlled by PC_3. When $PC_3 =1$, turn on the lamp, the lamp is turned off when the light is lit and $PC_3 =0$, which can be implemented as follows:

```
IN AL, 62H          ; Read the status of the C port
OR AL,00001000B     ; The D3 bit is set to 1 to light the light
OUT 62H,AL          ; Reset the C port status
```

The three instruction above is the function of the PC_3 set to "1", but does not affect the status of other bits of the port C. If you want to turn off the light, it will be "OR AL 00001000B" command into "AND AL, 11110111B", D_3 bit is cleared to 0. For the above operation through the C port according to the position/reset control word to achieve more intuitive.

The port C is in the position/reset control word format as shown in Figure 7-7.

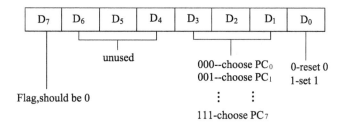

Figure 7-7 **C port by position/reset control word**

The 8 encoding of the $D_3 \sim D_1$ corresponds to the $PC_0 \sim PC_7$ pin, which is determined by the D_0 bit. The output of the bit is determined by 8255A, which is implemented by the position/reset control word, and the PC_3 is programmed as follows:

```
MOV AL,00000111B      ; PC3 sets 1
OUT 63H,AL            ; Output from control port
```

This method is more convenient and intuitive. It will take a position of port C "1" or "0" without affecting other bits of port C.

When the port A for mode 1 and mode 2 or 1 way for port B, C port the requisition of bits, the signal is generated by the hardware to decide, can not use C port according to the position/reset control word to be changed, but the interrupt enable flip-flop INTE can be set through the C port by location/reset control word. C port not requisitioned those bit the output mode, the control word is still valid.

7.1.5 Application of 8255A

Note that the port C is not written to the port C by the position/reset control word, but is written to the control port and is only valid for bits that are not taken in the port C and are set to output mode.

[**Example 7-2**] Assume that the A input in mode 1, allowing interrupts, that is, PC_4 set. Port B output in mode 1, allowing interrupt, that is, PC_2 set. Initialization procedures are as follows:

```
MOV AL,0B4H      ; According to the known conditions, the method is
                   10110100B = 0B4H
OUT 63H,AL       ; set working way
MOV AL,09H
OUT 63H,AL       ; PC_4 set, the A port input allows interrupts
MOV AL,05H
OUT 63H,AL       ; PC_2 set, the B port output allows interrupts
```

1. Application of 8255A

8255A mode 0 application example.

[**Example 7-3**] Use 8255A port A way, 0 output and parallel port printer connected, the memory buffer BUFF character print output.

Parallel data transmission between the printer and the host through the parallel communication mode, the process is the host will print the data sent to the data line, and then send a strobe signal $\overline{STB}$, printer data to read, while the BUSY line is high, notify the host number. Then stop sending, processing inside the printer to read the data. After the $\overline{ACK}$, while the BUSY failure, the host can notice the next data. The printer working sequence is shown in Figure 7-8.

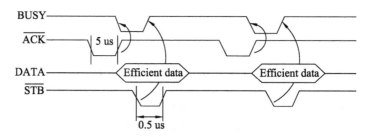

Figure 7-8 Working sequence diagram of printer

Design: according to the analysis of working characteristics of the printer, the printer needs a strobe signal latch start print to print the data through the BUSY signal, at the same time the printer state. In Figure 7-9, 8255A as CPU and printer interface, not only can work 8255A in $\overline{STB}$. Cases using the 0 query in 0 and needed to be defined C some of the bits as a contact signal and the 1 signal, BUSY connected. The two contact signal an output and an input, so C respectively on the high 4-bit and low 4-bit each define one (why). The definition of PC0 and $\overline{CS}$ cases.

CHAPTER 7   PROGRAMMABLE INTERFACE CHIP

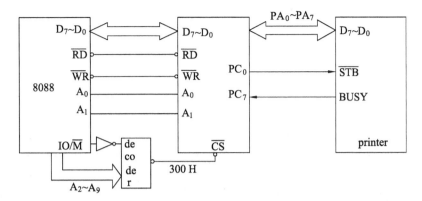

Figure 7-9   Diagram of connection between 8255A and CPU and printer

Note that in this case, PC_0 is an output control signal that is defined by itself, and its timing must be structured by software programming.

(1) Mode word setting

Port A mode 0, output; C mouth high 4-bit mode, 0 input, low 4-bit way, 0 output; port B is not used.

So the 8255A control word is:10001XX0B approach88H.

PC_0 set word: 00000001B approach 01H.

PC_0 reset word: 00000000B approach 00H.

(2) The programming

```
; Data segment definition
      BUFF DB 'This is a print program!',' $ '
; Code segment kernel
      MOV SI, OFFSET BUFF
      MOV AL, 88H          ; 8255A initialization, A port 0, output
      MOV DX, 303H
      OUT DX, AL           ; C port mode 0 input, low mode 0 output
      MOV AL, 01H
      OUT DX, AL           ; Causes PC₀ to be set, even if the strobe is
                           ;   invalid
WAIT:MOV DX,302H
      IN AL, DX
      TEST AL, 80H         ; Check whether PC₇ is 1, that is, busy
      JNZ WAIT             ; Wait for the sake of hurry
      MOV AL,[SI]          ; Print one character
      CMP AL,'$'           ; Terminator
      JZ DONE              ; Is the terminator, then the program is over
      MOV DX,300H
      OUT DX,AL            ; Not the terminator, but output from the A port
      MOV AL,00H
      MOV DX,303H
      OUT DX,AL            ; Make PC₀ clear 0, even if strobe is valid
      MOV AL,01H           ; Set PC₀ to 1OUT
```

```
            DX,AL              ; Produces a strobe signal, a negative pulse of a
                                 certain width
            INC SI             ; Modify the pointer to the next character
            JMP WAIT
    DONE:                      ; end of program
```

[**Example 7-4**] Instead of "8255A" as the keyboard and processor interface, write trip reverse method, keyboard scanning subroutine.

In order to identify the closed key on the keyboard, usually used in two ways: the line scanning method and line inversion method. A line scanning method has been mentioned in the Chapter 6, when the scan button is pressed on the keyboard, to go through multiple scans to find keys for low efficiency. It can fast inversion method identification key.

Figure 7-10 is a keyboard interface of 16 keys, 16 keys are arranged in a matrix of 4 x 4 columns, 4 lines from the output port of the A matrix $PA_3 \sim PA_0$, 4 column line is connected to the input port of B PB_3 to PB_0. with 16 keys to 16 hexadecimal number $0 \sim 9$ and $A \sim F$. each key rows and columns in a byte, 4 bit for high value for the low 4 column values, consisting of a TABLE, you can get the key value by looking up ranks value table. The definition of TABLE:

```
    TABLE DB 77H,7BH,7DH,7EH,0B7H,0BBH,0BDH,0BEH;0 ~ 7 scan code
          DB 0D7H,0DBH,0DDH,0DEH,0E7H,0EBH,0EDH,0EEH;8 ~ F scan code
```

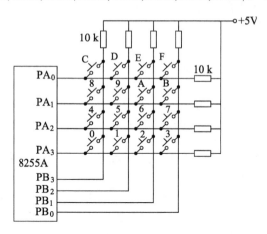

Figure 7-10 Keyboard arrangement, wiring and interface circuit

The procedure for identifying keys by line inversion is as follows:

① Set behavior output as input.

② Line output is 0, column value is read.

③ If the column value is all 1, it means no key to press, turn the step; if the column value is not all 1, it means that the key is pressed and the steps are executed.

④ Delay, eliminate jitter.

⑤ Read column values again, and column values, if they are all 1, indicate no key to press, turn the step; if the column value is not all 1, it is true that there is a key press, and execute step ⑥.

⑥ Set as output, behavior input.

⑦ Output the value of the non full 1 read just now from the column, and read the row value.

⑧ The column scan code is obtained, and the table lookup is converted to the key value.

⑨ After the button is released, the program is finished.

Assuming that the 8255A port A, B, C, and control port addresses are 300H ~ 303H. line inversion methods, the scan subroutine stores the key values of the key to the AH. The main code is as follows:

```
KEYSCAN PROC
SCAN: MOV DX,303H          ; Pointing control port
      MOV AL,10000010B     ; A port mode 0 output, B mouth mode 0 input
      OUT DX,AL            ; Write control word
      MOV DX,300H          ; A port address
      MOV AL,00H
      OUT DX,AL            ; To the A mouth, you output 0
      MOV DX,301H
WAIT: IN AL,DX             ; The keyboard status reads into the B port
      AND AL,0FH           ; Check only low 4 bit (column value)
      CMP AL,0FH           ; Are they all 1
      JE WAIT              ; No key press, continue to check the keyboard
; There is key down, delay 20MS, eliminate jitter, code slightly
      IN AL,DX             ; The keyboard status reads into the B port
      AND AL,0FH           ; Check only low 4 bit (column value)
      CMP AL,0FH           ; Are they all 1
      JE WAIT              ; Column value is all "1", no key press, continue
                           ;   to check the keyboard
      MOV KEY,AL           ; Key is pressed and column values are saved in the
                           ;   KEY unit
      MOV DX,303H          ; Set the 8255A control port address to change the
                           ;   way 8255A works
      MOV AL,90H           ; 8255A mode control word, B mouth way 0 output, A
                           ;   mouth mode 0 input
      OUT DX,AL            ; Output 8255A mode control word
      MOV DX,301H          ; B port address sent to DX
      MOV AL,KEY           ; Take out column values from the KEY unit
      OUT DX,AL            ; Output column values to the B port, reverse scan
      MOV DX,300H          ; A port address sent to DX
      IN, AL, DX           ; read line values from the A port
      AND, AL, 0FH         ; keep low 4 bits
      CMP AL,0FH
      JE SCAN              ; row value all "1", no key press, re scanning
      MOV, CL, 4, with keys pressed to form a row scan code
      SHL, AL, CL          ; row value left shift 4
      ADD, AL, KEY         ; component row scan code is deposited into the AL
                           ;   unit
```
The high 4-bit is the row value, and the lower 4-bit is the column value

```
        CALL KEYVALUE       ; check the TABLE table by column scan code in AL
        The key values are stored in AH, and the subroutine is abbreviated
        MOV, DX, 301H        ; B port address sent to DX, determine the key
                               release
        MOV AL,0
        OUT, DX, AL          ; output 0 to the B port, reverse scan
        MOV, DX, 300H        ; A port address to DX
        WAIT2:, IN, AL, DX   ; read line values from the A port
        AND, AL, 0FH         ; keep low 4 bits
        CMP AL, 0FH
        JNE WAIT2            ; the row value is not all "1", the key is not
                               released, and continue to wait for the release
                               of the key
        RET
    KEYSCAN ENDP
```

2. 8255A mode 1 application example

[**Example 7-5**] 8255A works in mode 1, where the string in the BUFF start buffer is printed from the printer with the $'end of the string. As shown in Figure 7-11, it uses 8086 8255A and 8259A processor by $\overline{CS}$ address decoder, the base address is 10H according to decoding figure, and 8259A base address is 20H. 8255A A0 connected to the 8086 address bus A2 and A1, so 8255A 4 port address: 10H, 12H, 14H, 16H. 8259A port address for 20H and 22H. When 8255A sends the data to the printer, PC_7 ($\overline{OBF}$) from high level to low level, from the falling edge of the monostable trigger can be generated by the printer strobe signal $\overline{STB}$, printer $\overline{ACK}$ issued after the end signal data as 8255A A port $\overline{ACK}$(PC_6), the signal can be set to PC_7 ($\overline{OBF}$) to become invalid, high level INTR (PC_3), also can make the signal effectively, IR3. 8259A issued a slightly connection with the system bus interrupt request signal connected to the system interrupt controller 8259A.

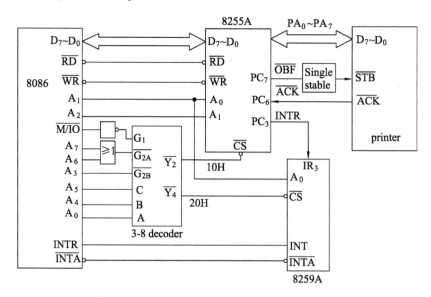

Figure 7-11 8255A mode 1 hardware connection

CHAPTER 7 PROGRAMMABLE INTERFACE CHIP

(1) Query mode transmission

Query mode, 8255A PC_7 ($\overline{OBF}$) is the state signal. when $PC_7 = 1$, and the output buffer is empty, CPU can output a 8255A to send data to the printer, the output data through the monostable circuit can automatically generate the printer $\overline{STB}$ signal, does not need to be like that cases of 7-3 through the C port set/reset $\overline{STB}$ word sequence.

The core program is as follows:

```
        MOV SI,OFFSET BUFF
        MOV AL,0A0H          ; 8255A mode word: 1010XXXXB.A port 1, output
        OUT 16H,AL
WAIT:   IN AL,14H            ; read C status
        TEST AL,80H          ; check whether PC₇(OBF) is 1, that is, busy
        JZ WAIT              ; PC₇=0,The output buffer is full while waiting
        MOV AL,[SI]          ; PC₇=1 Take a character output
        CMP AL,'$'           ; terminator
        JZ DONE              ; is the terminator, the end of the program
        OUT 10H,AL           ; not the end, the output from the A port,
                             ; automatic generation of
        INC SI
        JMP WAIT             ; prints the next character
DONE:   ...
```

(2) Interrupt mode transmission

8255A works in interrupt mode. Initialization includes the setting of the 8255A control word and the setting of the interrupt enable bit $INTE_A$.

The 8255A control word is "1010XXXX B", that is, "0A0H."

The setting of $INTE_A$ is realized by PC_6 setting, and the port C setting word is 00001101B, that is 0DH.

Assuming that 8259A is 08H when initialization is ICW_2, then the interrupt type code for the 8255A port A is 0BH, and the interrupt vector corresponding to this interrupt type code should be placed in the 4 cell of the interrupt vector table starting from 2CH.

In order to accurately find the character to be printed in the interrupt service program in the data segment defines the buffer pointer is used to access the POINT to print characters, also set a DONE symbol to synchronize with main program. Also note that, the first character must be started in the main program in programming print, subsequent characters can in order to interrupt the transmission. Assuming that 8259A is initialized, 8255A and interrupt the main code:

```
; Master program kernel code
MAIN:  MOV  AL,0A0H          ; 8255A mode word
       OUT, 16H, AL          ; set the control word for 8255A
       MOV, AL, 0DH          ; PC₆ is set to 1
       OUT 16H, AL           ; causes 8255A A port output to allow
                             ; interrupts
;Set the interrupt vector section to refer to the 7.4 section,
       MOV AX,SEG BUFF
```

```
            MOV DS,AX
            MOV POINT, OFFSET BUFF   ; sets the address pointer
            MOV DONE, 0              ; interrupt end flag bit, DONE = 1 all characters
                                       printed
            IN, AL, 22H              ; open 8259A interrupt mask bit, IMR3 = 0
            AND AL,11110111B
            OUT 22H,AL
            STI                      ; SETB EA
            INT 0BH                  ; print the first character by calling the
                                       interrupt service program through the
                                       interrupt instruction
     NEXT: CMP, DONE, 0              ; DONE = 0, the print is not over yet, and
                                       continue to wait for the print to end
            JZ NEXT
                                     ; INTEND:…All interrupts completed, interrupt
                                       end processing done
                                     ; interrupt service subroutine kernel code
     ROUTINTR PROC                   ; protect the scene slightly
            MOV, DI, POINT           ; remove the address of the characters
                                       to be printed, accessed indirectly
                                       through the DI
            INC POINT                ; address pointer plus 1
            MOV, AL, [DI]            ; remove the characters to be printed
            CMP AL,'$'
            JZ FINISH
            OUT 10H,AL               ; A character is output from the A port, and a
                                       printer strobe signal is produced
                                       simultaneously
            JMP EXIT
     FINISH:, MOV, DONE, 1           ; all characters are printed and done
                                       end of interrupt processing
            MOV, AL, 0CH             ; DONE = 1, synchronization with main program
            OUT, 16H, AL             ; prohibit 8255A A port output interrupt
            IN, AL, 22H              ; set the 8259A interrupt mask bit, IMR3 = 1
            OR AL,00001000B
            OUT 22H,AL
     EXIT:, MOV, AL, 20H             ; send 8259A's EOI command
            OUT, 20H, AL             ; output from 8259A's even address
            IRET                     ; restore the scene slightly
     ROUTINTR ENDP
```

Note: after the end of the interrupt, prohibit interrupting 8255A and set 8259A operation can be placed in the interrupt service program processing part of the end, can also be placed at the end of the main program interrupt processing part.

7.2 Serial communication and serial interface

Serial communication refers to the use of a data line, the data will be an orderly transfer, every data occupy a fixed length of time. Only a few lines will be able to exchange information between systems, especially suitable for long distance communication between computer and computer, computer and peripherals.

7.2.1 Mode of serial communication

The two sides of the communication can be divided into three different ways according to the direction of data transmission, as shown in Figure 7-12.

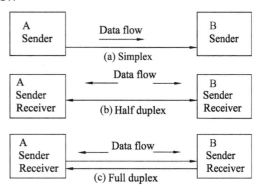

1. Simplex mode

Only data is allowed to be transmitted in a fixed direction, that is, a party can only serve as a transmitter station, and the other party can only be used as a receiving station

2. Half-duplex mode

Figure 7-12 Data mode

The data can be transmitted from the station A to the station B, and also from the station B to the station A, but cannot be transmitted in two directions at the same time. Only one station can be sent at a time and another station can receive the communication. Both sides can send and receive in turn

3. Full duplex

The two sides also allows communication to send and receive. At this time, A can also send in the station receiving, station B is the same. The full duplex mode is equivalent to two opposite simplex mode together, so it needs two transmission lines.

The main use of simplex and duplex mode in computer serial communication.

7.2.2 Serial communication classification

Serial communication, data, control and status information is to use the same transmission line, in order to make the communication process correctly and smoothly, send and receive both parties must abide by the common communication protocols, in order to solve the transmission rate, information format, synchronization is data verification and other issues. According to the synchronous mode, will be divided into two serial communication class: synchronous and asynchronous communication.

1. SYCOM

Synchronous communication is a way of sending and receiving data communication over a clock signal under the common control. In a cycle time synchronization clock signal, sending data to the sender of an online synchronization data. Synchronous communication requirements clock to maintain strict synchronization.

In synchronous communications, data is transmitted as frames, and a frame usually contains several data characters.

The information frame consists of a synchronous character data block and check character. The character in the beginning frame synchronization is used to confirm the data characters start. Data blocks in synchronous character, without limiting the number required by the transmission of the data block length is decided; check character has 1 to 2, to check the receiver correctly the character of the received sequence. Synchronous communication data format is shown in Figure 7-13.

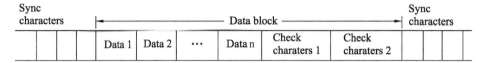

Figure 7-13 Data format for synchronous communication

Step communication can be divided into the following two types according to the type of data transmitted:

(1) Bit oriented (bit) type synchronous mode in binary bit as a unit of information. In synchronous transmission, each data block in the head and tail of a special sequence of bits (such as 01111110) to mark the beginning and end of the data block data block. As a bit stream to deal with, and not as a character stream in order to avoid data processing. In the special mode of marking the beginning and end of the block in current methods, bit insertion is usually used as the sender send data in the stream, after five consecutive "1", insert a 0. Receiver in the data stream is received, if detected five consecutive 1 sequence, to check a data later, if the bit is 0, it is removed; if the bit is 1, said the end of the data block, to the end. For a flow of the typical synchronous communication mode is mainly applied to computer network communication protocol, such as the high level data link control protocol (HDLC-High-Level Data Link Control).

(2) Character oriented type synchronous way: with the character as a unit of information. The character is a EBCD code or ASCII code. In synchronous transmission, the beginning of each data block in the head with one or more synchronous character SYN to mark the data block; the tail with another unique character mark the end of the ETX block. Any of these special characters, ordinary character bit patterns and transmission have different significant.

2. Asynchronous communication

The two sides do not need to send and receive asynchronous communication for data transmission control in the same clock signal, each by its own clock signal to send and receive data. Data transmission in character units, between characters and characters of transmission is asynchronous, it can have arbitrary time intervals, synchronous transmission with in characters.

Data are usually in character or byte characters frame transmission. Transmission of a character in the serial asynchronous communication information in the format prescribed start bit, data bits, parity bits, stop bits, including all the significance of Figure 7-14 asynchronous communication data format shown in:

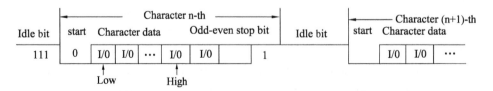

Figure 7-14 Data format for asynchronous communication

(1) Start bit: first sends a logical "0" signal to indicate the beginning of the transmission character.

(2) Data bit: following the start bit, the number of data bits can be 5,6,7,8 and so on, forming a character. The ASCII code is usually used. It starts from the lowest bit and is positioned by the clock.

(3) Parity bit: after this bit is added to the data bit, the number of "1" should be even (even checked) or odd (odd check) to verify the correctness of the data transfer.

(4) Stop bit: it is the end of a character data. It can be 1-bit, 1.5-bit, 2-bit high level.

(5) Idle bit: in the logical "1" state, which indicates no data transfer on the current line.

Character frame by frame sent by the sender, the transmission line is the receiving device frame is received. The low level logic receiver detects the transmission line to send over 0″ (i. e. the character frame start bit), determine the sender has begun to send data, whenever received stop character frame when it know a frame character has been sent. Any time the interval between characters and characters. The receiving device after receiving the initial signal transmission time and as long as the sending device of a character to keep pace can be received correctly, the arrival of the next character start bit and synchronizing the recalibration.

Assuming the transmission side transmitting character "9", namely ASCII 39H, asynchronous communication data format for a start bit, 7 data bits, a parity bit and 1 stop bit. According to characters from low to high order sent, then transferred from the signal line of the data flow is as follows: 0100111001.

7.2.3 Rate of serial communication

Before and after data transmission, both receiver and receiver must unify the data transmission mode of both sides, and also unify the speed of data sending and receiving on both sides

1. Data-signalling rate

Data transfer rate refers to the amount of information transmitted in unit time, and can be expressed by bit rate and baud rate.

(1) Bit rate: bit rate is the number of binary bits transmitted per second, expressed in BPS (bit/s).

(2) Baud rate, baud rate is the number of symbols transmitted per second.

Sometimes the baud rate in conjunction with the bit rate of confusion, in fact the latter is on the information transmission rate (transmission rate) measurement. The baud rate can be understood as a unit of time transmission symbol number (transmission symbol rate), through the different modulation methods can load a plurality of bits of information in a symbol. It uses per unit time. Carrier modulation state changes number, the unit is Potter (Baud). Obviously, two-phase modulation (corresponding to a single modulation state of 1-bits) of the bit rate is equal to the baud rate; four phase modulation (corresponding to a single 2-bit hexadecimal two state modulation) bit rate is two times the baud rate of eight (single phase modulation; modulation The bit rate of the state corresponding to the 3 binary bits is three times the baud rate; in turn, the relation between the baud rate and the bit rate is bit rate = baud rate * the number of binary bits corresponding to a single modulation state.

In the computer, a symbolic meaning to the high level, they represent the logic "1" and

"0", so the amount of information contained in each symbol is 1 bit, so in computer communication, often referred as the baud rate of the bit rate, i. e. :

1Baud(baud) = 1bit/s

For example: the fastest teletype transmission rate is 10 characters per second, each character contains 11 bits, then the data transmission rate: 11bit/ character * 10 character /s = 110bit/s = 110Baud. Commonly used baud rates are as followed in computer: 110, 300, 600, 1200, 2400, 4800, 9600, 19200, 28800, 33600, 115200, 56 ×10^3. (The units are bit/s)。

For example: receive and send communication according to the asynchronous mode of communication data transfer rate of 120 /s characters, each character consists of 7 data bits, 1 bit parity and 1 stop bit. Then each character frame 10 (1 initial +7 bit +1 bit data check the + 1 bit stop bit), the transmission bit rate of 10 bit/ * 120 /s = 1200 bit/s.

2. Baud rate factor

Serial communication process, in order to ensure communication can both send and receive reliable data, usually send a n sending and receiving periodic clock and receiving a message. At this time, as the reception clock and the transmission clock and baud rate: $F = n * B$. where F is sending or receiving the clock B is the data transmission frequency; baud rate; n is called the baud rate factor. In the actual serial communication, baud rate factor can be set.

For example, in accordance with the sending and receiving data transmission speed per second 9600 Potter, baud rate factor is 16, then every 16 clock time to transfer one bit of information, then send and receive clock clock frequency is $F = 16 * 9600 = 153600 Hz$.

7.2.4 Serial interface standard RS-232C

In order to realize the computer from different manufacturers and various external equipment serial communications, developed a number of international standard serial interface, common RS-232C interface, RS-422A interface, RS-485 interface. At present, the most commonly used is the RS-232C interface standard RS-232C interface standard of the U. S. electronics industry association issued the provisions of the mechanical, electrical parameters, function etc. hand. It as a standard, widely used in computer communication interface.

The RS-232C standard was originally connected with remote communication data terminal equipment DTE (Data Terminal Equipment) and data communication equipment DCE (Data Communication Equipment). But it has been widely used in computer borrowed (more precisely, a computer interface) and the terminal or between peripherals attached to the proximal end of the. RS-232C standard in the "send" and "receive", is standing in the DTE position, rather than standing in the position of DCE in computer to the definition. In a system, information is often sent between CPU and I/O devices, both of which are DTE, so that both can send and receive.

1. Signal level

RS-232C uses negative logic representation of data, logic "1" level for 3 ~ 15V, logic "0" level for 3 ~ 15V. In actual use, often using ±12V or ±15V. It shows that widely adopted signal TTL in computer is not compatible with signal level in standard RS-232C when in use, there must be a level conversion circuit.

2. SPD

Serial interface standard refers to a computer or terminal (data terminal equipment DTE) serial interface circuit and modem MODEM (data communication equipment DCE) connected between the standard. Due to the physical characteristics of RS-232C, it does not define the connector thus emerged as the connector shown in Figure 7-15 of the DB-25 and DB-9 types, the definition each of pin also is not identical, but most devices are using DB-9 type connector pins are defined as follows:

(1) T_XD: send data line, output. Send data to MODEM.

(2) R_XD: receive data line, input. Receive data to computer or terminal.

(3) RTS: request send, output. The computer informs the MODEM through this pin and requests data to be sent.

(4) CTS: allow send, enter. Send CTS as an answer to RTS, the computer can send the data.

(5) DSR: the data device is ready (that is, the MODEM is ready). The input indicates that the modem can be used and that the signal is sometimes directly connected to the power supply so that the device is valid when connected.

(6) CD: carrier sense (receiving line signal detector), input. It indicating that MODEM is connected to the telephone line.

(7) GND: ground. If a communication line is part of an exchange telephone, two signals at least are needed:

① RI: ringing instructions, enter. MODEM, send the signal to notify the computer or terminal if it receives a ringing call from the switchboard.

② DTR: the data terminal is ready and output. After the computer receives the RI signal, it sends the DTR signal to the MODEM as an answer to control its conversion device and establish a communication link.

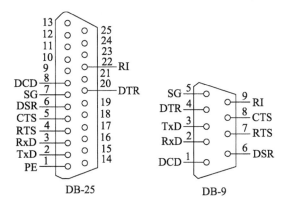

Figure 7-15 DB25 and DB9 connector interface diagram

3. RS-232C connection

(1) Using MODEM for connections

If a telephone line is used for communication between microcomputers, then the computer must be connected to the MODEM via the 232C interface, and the MODEM is connected by telephone lines. The connection and communication principle is shown in Figure 7-16.

Figure 7-16 The microcomputer connects through MODEM and communication principle

The 232C interface of the computer and the 232C interface of the MODEM are connected to the 25 pin MODEM connecting line through the 9 pin serial port. The 9 pin female plug connected with the computer is connected with the 25 pin male plug of the MODEM. The correspondence between signals, such as Table 7-2, is shown in the wiring correspondence table

Table 7-2 Connection correspondence table

signal	9needle	25needle
TXD	3	2
RXD	2	3
RTS	7	4
CTS	8	5
DSR	6	6
GND	5	7
CD	1	8
DTR	4	20
RI	9	22

(2) Two direct connection between communication devices

If you have a serial communication between two close computers or between computer and peripheral, you can connect directly between two devices without MODEM, and there are 3 specific ways.

① The most simple three wire connection, as shown in Figure 7-17(a), using only RXD, TXD and GND connected to the three signal. The procedure does not need to make the RTS and DTR, CTS and DSR have to judge whether it is effective. This method is simple, economic, but not reliable.

② The two devices still use three lines, but the interface of RTS and CTS respectively DTR and DSR each interconnection, interconnection, as shown in Figure 7-17(b) shown in the program. The RTS and DTR, indicate that the request transfer device is always allowed, data is always ready, the two sides at any time for serial receive, send, this way is called a pseudo contact three wire connection. By this way the communication connection may not be reliable.

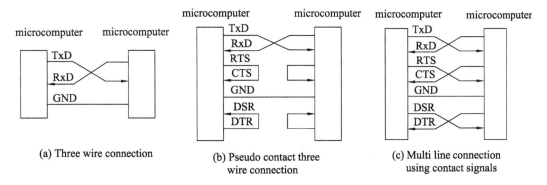

(a) Three wire connection (b) Pseudo contact three wire connection (c) Multi line connection using contact signals

Figure 7-17 A direct connection between two devices

③ The sending and receiving the most reliable connection is shown in Figure 7-17(c) shown by the use of contact signal multi line type connections. In this way, only the signal to both CTS and DSR can receive effective RTS and DTR for data transmission. When the receiver to processing the received data wait, can make their own procedures RTS signal invalid, making the sender CTS invalid, the sender sends the data receiving end suspended; processing the received data, set the RTS signal, both sides may continue to send data. This is the working principle of hardware flow control.

Note that before the two connection are not reliable, whether the two sides can not confirm the normal communication through hardware. Both sides of communication before the data transmission to the best software handshake to confirm whether the communication is reliable. For example, the sender first sends a specific character, the receiver receives the character after the return to the sender to confirm the information, only the sender receives sure the information is correct, only shows that the connection is reliable. After that both sides can communicate normally.

7.3 Programmable serial interface 8251A

8251A is a programmable serial communication interface chip. It can work in synchronous mode and asynchronous mode. Under synchronous mode, the baud rate is 64 Kbit/s, under asynchronous mode, the baud rate is 0 ~ 19.2Kbit/s.

In synchronous mode, each character can be represented by 5,6,7 or 8 bits, and the inner can automatically detect synchronous characters to achieve synchronization. In addition, 8251A also allows the odd/even parity bits to be checked in synchronous mode.

Asynchronous mode of each character can also be expressed by 5,6,7 or 8, the clock frequency of the baud rate of 1,16 or 64 times, with 1 as the odd/even parity,1 start. And according to the programming data for each increase of 1, 1.5 or 2 stop bits. In asynchronous mode, 8251A can check the start, automaticly detect and treat of the terminating character.

7.3.1 The internal structure of 8251A

The internal 8251A is composed of five parts: data bus buffer, read/write control circuit, transmitter, receiver and modulation demodulation control circuit. The internal structure is shown in Figure 7-18, and the specific functions are as follows:

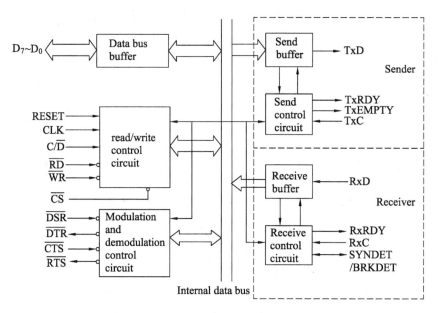

Figure 7-18 Internal structure diagram of 8251A

1. The data bus buffer

The data bus buffer is the data interface between the CPU and the. CPU, the 8251A gives the initialization command word of the 8251A, the status information of the 8251A, and the input/output data is exchanged with the CPU through the data bus buffer.

2. The read/write control circuit

The read/write control circuit cooperates with the data bus buffer to complete the initialization of the CPU, the reading of the status information, and the input and output of the data of the 8251A.

3. The transmitter

The transmitter is composed of two parts, the sending buffer and the transmitting control circuit.

The transmitter receives the parallel data from CPU, save it in the transmission buffer, then 8251A according to the different ways of working in controlling transmission control circuit, the parallel data into serial data output from the serial data line T_XD in accordance with the provisions of the data format.

4. The acceptor

The receiver consists of two parts, the receiving buffer and the receiving control circuit.

The receive buffer receives the serial data from the R_XD pin and, under the control of the receiving control circuit, converts the serial data into parallel data and then stores the received buffer in accordance with different operation modes.

5. Modulation and demodulation control circuit

The modulation and demodulation control circuit are used to control the connection between the 8251A and the modem.

7.3.2 8251A pin function

The 8251A is a 28 pin dual in-line chip with pins such as Figure 7-19, pin signals fall into three classes.

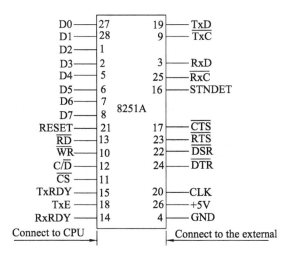

Figure 7-19   Pin diagram of 8251A

1. The connection signal between 8251A and CPU

(1) $\overline{CS}$: The chip select pin, input, active low. It consists of the address signal by CPU after decoding.

(2) $D_0 - D_7$: The 8-bit, three state, bidirectional data pins are connected to the data bus of the system. Transmit CPU to the 8251A programming command word and 8251A to the status information and data of CPU.

(3) $\overline{RD}$: Read the control pin, input. For low power, the CPU is reading data or status information from the 8251A.

(4) $\overline{WR}$: Write the control pin, input. For low power, the CPU is writing data or controlling information to the 8251A.

(5) C/$\overline{D}$: control/data selection pin input. To distinguish the current read/write data or control information and state information of. C/$\overline{D}$ is low, said to read and write data ports; C/$\overline{D}$ is in high level, said to read and write to the control port.

The combination of $\overline{RD}$, $\overline{WR}$, C/$\overline{D}$ these 3 signals determines the specific operation of the 8251A, as shown in Table 7-3.

Table 7-3   Relationship between port read and write operation and pin signal

C/$\overline{D}$	$\overline{RD}$	$\overline{WR}$	operate
0	0	1	Read data input port (receive)
0	1	0	Write data output port (send)
1	0	1	Read status information
1	1	0	Write initialization command word

(6) RESET: The reset pin, when the high level. RESET signal is valid, the 8251A receive and receive lines are in idle state and wait for CPU to initialize its programming.

(7) T_xRDY: The sender prepares the pins. The output is used to notify the CPU that the

8251A is ready to send a character. In the interrupt mode, TXRDY can be used as the interrupt request signal, and TXRDY is used as the query signal in the query mode

(8) T_XE: The transmitter null pin outputs. TXE valid for high power. This is used to indicate that the 8251A transmitter is parallel to the serial converter blank, indicating that a sending operation has been completed.

(9) R_XRDY: The receiver is ready for pin output. 8251A has been used to indicate the current from an external device or modem receives a character, waiting for CPU to take. Therefore, in interrupt mode, RXRDY can be used as the interrupt request signal; in the query mode, RXRDY is used as the query signal.

2. Connecting signals between 8251A and external devices

The connection signals between the 8251A and the external devices are divided into two categories: data signals and receive and receive signals, as shown in Figure 7-19.

(1) T_XD: transmitter data output signal. When CPU parallel data sent to 8251A is converted to serial data, it is sent to peripheral via TXD.

(2) R_XD: receiver data input signal. It is used to receive serial data sent by peripheral, and the data is converted into parallel mode after entering 8251A.

(3) SYNDET: the synchronization detection signal. The signal is only used for two-way synchronization. Synchronization, when 8251A synchronous signal is detected, the output signal to the SYNDET MODEM that has been used to external synchronization; synchronization, synchronous signal, the synchronous receiving MODEM after receiving data information.

$\overline{DTR}$: The data terminal is ready for the signal, output, active low. Used to notify the external device that CPU is currently ready.

$\overline{DSR}$: The data device is ready for the signal, input, active low. Indicates that the current peripheral is ready.

$\overline{RTS}$: Request a send signal, output, active low. Indicates that the CPU is ready to send data.

$\overline{CTS}$: Enable send signal, input, active low, response to $\overline{RTS}$, sent from peripheral to 8251A. 1

Note: $\overline{DTR}$ and $\overline{DSR}$ are a pair of contact signals. $\overline{RTS}$ and $\overline{CTS}$ are a pair of contact signals. If 8251A does not communicate with the outside world directly without using a modem, you can use the two pairs of contact signals, but you must connect the 8251A $\overline{CTS}$ to the ground.

3. Clock, power and ground

In addition to the CPU and peripheral signals, the 8251A also has 3 clock signals and power and ground signals.

$\overline{TxC}$: Transmitter clock, input is used to control the speed. In send characters the falling edge of the data from the 8251A output shift. Synchronous mode, $\overline{TxC}$ frequency is equal to the character transmission baud rate; asynchronous mode, the $\overline{TxC}$ frequency is $\overline{TxC}$ times the baud rate of the character, 16 or 64 times.

$\overline{RxC}$: The receiver clock, which is used to control the speed of the received character, collects the serial data input line at the rising edge of the $\overline{RxC}$. The frequency requirement is the same as that of the $\overline{\text{TxC}}$.

CLK: The clock input is used to generate the internal timing of the 8251A. The frequency of the. CLK should be at least 30 times the receive clock ($\overline{\text{RxC}}$ and $\overline{\text{TxC}}$), the transmit clock frequency (synchronous mode), or 4.5 times (asynchronous mode).

In actual use, $\overline{\text{RxC}}$ and $\overline{\text{TxC}}$ are often linked together, provided by the same external clock, and the CLK is supplied by another higher frequency external clock.

V_{CC}, GND: power and ground input signal. V_{CC}, generally connects to +5V.

7.3.3 8251A operate mode

1. Sending data

When the transmitter is ready, the TXRDY pin becomes high, indicating that the CPU can send data to the transmitter.

(1) Asynchronous transmission

The transmit control circuit is coupled with the start bit, the parity bit and the stop bit at the beginning and the end of the information to be transmitted, and then starts from the start bit and is shifted bit by bit from the data output line TxD to bits by bit.

(2) Send synchronously

Before sending data, the transmitter will automatically send 1 or 2 synchronous characters serially, and then output data by bit. The serial output check character after the whole data transfer is finished.

2. Data reception

(1) Asynchronous receive

The receiver first detects the start bit, the process is shown in Figure 7-20. When the receiver detects the R_xD line low level after the start of a counter, and 1 each receive clock counter, when the counter counts to n/2 (n baud rate factor, n/2 said the middle point, a baud rate factor graph is 16), again sampling R_xD signal line, if the R_xD signal line is still low, then confirm receipt of an initial information frame. If the signal sampling is of high level and low level said just now is the interference signal, to detect R_xD line.

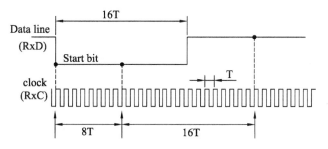

Figure 7-20 Detection of start position

The detection of the start bit, 8251A every N receive clock to R_xD sampling time, the received signal is data bits, parity bit and stop bit, parity checks in accordance with the provisions of the format and remove the stop position, consisting of parallel data, to the data input register, issued at the same time, the R_xRDY signal sent to CPU, said it had received a available data.

(2) simultaneous reception

The receiver first searches character synchronization monitoring. 8251A R_xD line, R_xD line whenever one data bit is received, and down to the shift register shift, compared with synchronous character register contents, if they are not equal, it receives a data, and repeat the above process. When the two register contents are equal 8251A, SYNDET rose to high level, said synchronous character has been found, synchronization has been achieved.

If the two synchronization mode, receiving the first sync character, then continue to test whether the contents of the input shift register and second synchronous character register the same content. If the same, that synchronization has been achieved. If not the same, then retest the first synchronization characters.

In the case of external synchronization, the synchronous character is detected by an external MODEM, and when the MODEM detects the synchronization information, it sends the SYNDET signal to the 8251A to indicate that it has been synchronized.

After synchronization, the receiver samples the R_xD line with the receive clock signal and sends the received data to the shift register. After receiving a character, a high-level signal is sent on the R_xRDY pin.

7.3.4　8251A internal registers and initial programming

8251A is a programmable multi-function serial communication interface chip. It must be initialized before it is actually used. It is used to determine its working mode, transmission rate, character format and stop bit length.

8251 internal mode control word register command controls word register, state register and register with the odd address synchronous character. ($C/\overline{D}=1$) access. There are two data registers: the data input register and data output register, with even address ($C/\overline{D}=0$) access.

1. Mode control word (mode word)

The mode control word format is shown in Figure 7-21. It is used to determine the 8251A mode of operation, the transmission rate, the character format, and the stop bit length.

(1) D_1D_0: to determine the work in synchronously or asynchronously. When $D_1D_0=00$ is synchronous mode; when the asynchronous mode is $D_1D_0=00$, three combinations and D_1D_0 in order to select the input proportion coefficient between the clock frequency and the baud rate.

(2) D_3D_2: to determine the number of digits contained in 1 data.

(3) D_5D_4: to determine whether or not to check and verify the nature of parity.

(4) D_7D_6: The meaning of synchronous and asynchronous methods are different. The number of bits used to specify a stop bit is used asynchronously. The synchronization is used to determine whether it is internal or external synchronization, and the number of synchronization characters.

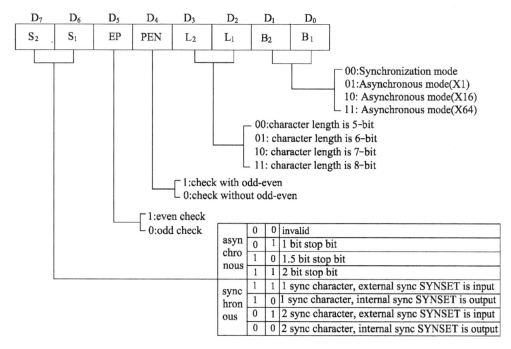

Figure 7-21 8251A mode selects the format of the control word

[**Example 7-6**] An asynchronous communication, the data port address for the 300H control port address for 301H (the same below), the data format using 8 data bits, 1 start bit, 2 stop bit, parity, baud rate coefficient is 16, the working mode of word 11011110B =0DEH.

```
MOV DX,301H       ; 8251 command port address
MOV, AL, 0DEH     ; asynchronous working word
OUT DX,AL
```

[**Example 7-7**] In synchronous communication, if the frame data format is: character length is 8 bits, double synchronization character, internal synchronization mode, odd check, its working word is 00011100B =1CH.

```
MOV, DX, 301H     ; 8251 command port
MOV, AL, 1CH      ; synchronous working word
OUT DX,AL
```

2. Operation command control word (control word)

The command control word is used to control the operation and status of the 8251A, forcing the 8251A to perform some operation or to work in order to receive or send data. The. 8251A operation commands the format of the control word, as shown in Figure 7-22.

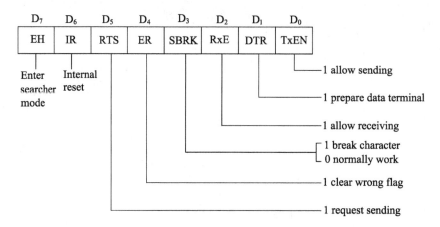

Figure 7-22 8251A The format of the operation command control word

(1) D_0: allow TxEN (Transmit Enable) to send data enable bit. $D_0 = 1$, allow send; $D_0 = 0$, prohibit sending.

(2) D_1: the data terminal is ready for DTR. $D_1 = 1$, setting the $\overline{DTR}$ pin output to a valid low level, indicating that the terminal device is ready; $D_1 = 0$, setting $\overline{DTR}$ null and void.

(3) D_2: allow to receive RxE (Receive Enable), receive enable bit. $D_2 = 1$, allow receive; $D_2 = 0$, prohibit receiving.

(4) D_3: send SBRK (Send Break Character termination character). $D_3 = 1$, forcing TxD to a low level, the output of the continuous space.

(5) D_4: clear the error flag and clear all the error flags in the status register 0.

(6) D_5: request to send signal RTS. $D_5 = 1$, $\overline{RTS}$ pin output active low, said terminal equipment is ready for the D5 = 0, the $\overline{RTS}$ invalid.

(7) D_6: internal reset IR (Inter Reset). $D_6 = 1$, so that 8251A back to select the command state mode; $D_6 = 0$, do not go back to mode command.

(8) D_7: enter the search mode EH (Enter Hunt Mode). In synchronous mode, $D_7 = 1$ starts the search synchronization character; $D_7 = 0$ does not search for synchronous characters. Under asynchronous mode, the bit is invalid.

[**Example 7-8**] To make the 8251A $\overline{RTS}$ output valid level acceptable and allow to be sent, the program segment that sets the operation command word is:

```
MOV, DX, 301H      ; 8251A command port address
MOV AL, 00100101B  ; sets D5, D2, and D0 1, so that RTS is valid, allowing
                     receiving and sending
OUT DX,AL
```

3. Status word

Status word 8251A to execute a command after the data transfer on status word register, CPU can read 8251A word, analyze and judge, to decide how to do the next step of the. 8251A word format as shown in Figure 7-23 (all state is set "1" valid).

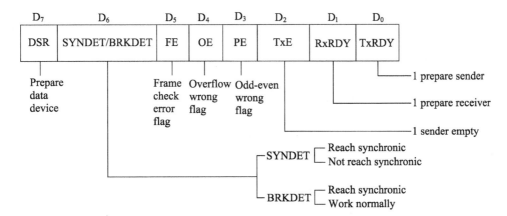

Figure 7-23 8251A Format of status word

The status register state RxRDY, TxE, SYNEDET and DSR defined the definition and pin the same, only the meaning of TxRDY 8251A chip pin on the meaning of TxRDY is different. The state TxRDY status register, as long as the empty set a send buffer; and pin TxRDY to meet $\overline{CTS}=0$ and TXE =1. To meet the three conditions before setting the. $D_3 \sim D_5$ is three bit error status information:

D_3: parity error PE (Parity Error). When parity is detected by the receiver, PE sets "1". PE valid and does not prohibit 8251A work. It is reset by the ER bit in the work command word.

D_4: overflow error OE (Overun Error). If the first character of CPU has not been removed, after a character has become effective, OE is set to "1". OE does not prohibit the operation of 8251A, but the overflow character lost, OE was working command word ER reset.

D_5: frame error FE (Frame Error, used only for asynchronous). If the receiver does not detect a specified stop bit after the one character, the FE sets "1". The ER is reset by the working command word, which does not affect the operation of the 8251A.

[**Example 7-9**] To query that the 8251A receiver is ready, use the following program segment:

```
      MOV DX,301H        ; 8251A command port
L:IN, AL, DX, read status word
      AND, AL, 02H       ; check D₁ =1? (RxRDY =1?)
      JZ L               ; unprepared, waiting
      MOV, DX, 300H      ; data port
      IN, AL, DX; ready, then reading
```

4. Synchronous character register

8251A has two 8-bit synchronous characters registers, used to save two characters synchronous synchronous communication. The number of synchronous characters is controlled by the mode before sending synchronization 8251A first sends the first synchronous character of the register, if there are two synchronization characters, is followed by the synchronization of characters sent second synchronous character in the register, and then send the data. The serial receiver, synchronization character save synchronization characters received and synchronous character register in comparison to determine whether synchronization.

5. Initialization of 8251A

Model and control words of 8251A have no characteristic sign, use the same control port address, according to the order of 8251A written to distinguish, the word is written first mode, after the write controls word is, in order to avoid the mode control word written into other ports, 300H and 1 40H have been written to the control port before the initialization procedure, known as the internal reset command. 8251A initialization process is shown in Figure 7-24.

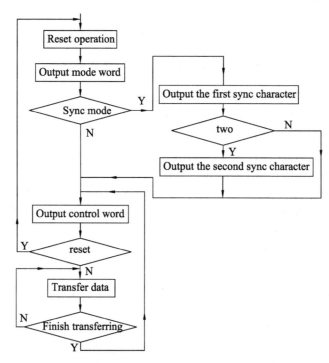

Figure 7-24 8251A Initialization flow chart

8251A work in asynchronous mode, initialization of just two steps, the first output mode control word (mode), and then output the control word. In synchronous mode, writing mode control word (mode), according to the number of synchronous character output one or two synchronous character, finally output the control word.

Note: 8251A is initialized to the same port for write operation, in order to allow 8251A to reliably initialize between successive OUT instructions should be added to the appropriate delay, you can use the NOP command, can be realized by the delay subroutine.

(1) Example of initialization program in asynchronous mode

[**Example 7-10**] 8251A mode is asynchronous mode and baud rate coefficient for 64, 5 data bits/character, parity, 2 stop bit, sending, receiving permits, a data port address is 300H, the control port address for 301H, write the initialization procedure.

According to the requirements of the topic, you can determine the mode word is: 11110011B, that is, F3H. The control word is "00110111B", that is "37H".

The initialization procedure is as follows:

```
        MOV AL,0F3H         ; Send mode word
        MOV DX,301H
        OUT, DX, AL         ; asynchronous mode, 5 bit/character, parity check, 2
                              stop bits
        NOP
        NOP
        MOV, AL, 37H        ; set the control word to send, receive, allow, clear the
                              error flag, and make the RTS and DTR pin output low
        OUT, DX, AL         ; valid
```

(2) Example of initialization program in synchronous mode

[**Example 7-11**] The 8251A data port and the control port address are 300H and 301H respectively. In the internal synchronization mode, 2 synchronous characters (set the synchronization character to 16H), parity check, 8-bit data bits/characters, and write the initialization program

According to the requirements of the topic, you can determine the mode word is: 00111100B, that is, 3CH.

The control word is: 10010111B 97H. According to the control word, we can know 8251A on synchronous character retrieval; 3 error flag is reset in the status register; 8251A sending, receiving permission; CPU is ready for data transmission。

The initialization procedure is as follows:

```
        MOV, AL, 3CH        ; set mode word, synchronous mode, use 2 synchronous
                              characters
        MOV DX,301H
        OUT, DX, AL         ; 7 data bits, even check
        NOP
        NOP
        MOV AL,16H
        OUT, DX, AL         ; send synchronous character 16H
        NOP
        NOP
        OUT DX,AL
        NOP
        NOP
        MOV AL, 97H         ; sets the control word to enable the transmitter and
                              receiver to start
        OUT DX,AL
```

7.3.5 Application of 8251A

1. Example of Programming with Status Words

[**Example 7-12**] Query from the peripheral serial inputs 10 characters, and the received characters stored in the buffer BUFF. 8251A connection is as shown in Figure 7-25.

As can be seen, 8251A data port and control port address are 300H and 301H, respectively. As can be seen, if 8251A works in asynchronous mode, the baud rate factor is 16, 7 data bits, 1 stop bits, parity, mode word requirements for 01011010B. 8251A works in full duplex mode, and effectively, eliminates error flag, controls word for 00010111B.

The input process in the query mode is as follows: first set the mode word, and then set the control word, the initial read status word, query RxRDY to 1, you can read data from the data port, and if the input error is reported wrong.

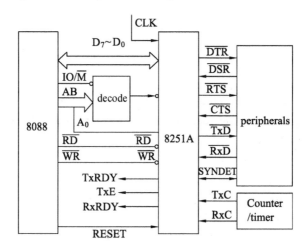

Figure 7-25 Serial communication connection diagram

The main program code is as follows:

```
         MOV CX, 3
         XOR AL, AL
         MOV DX, 301H
    AGA: OUT DX, AL
         CALL DELAY
         LOOP AGA
         MOV AL, 40H      ; Soft reset command
         OUT DX, AL
         CALL DELAY
         MOV AL, 5AH      ; Set mode word
         OUT DX, AL       ; 7 bit data, 1 bit stop bit, odd check, baud rate
                            factor 16
         CALL DELAY
         MOV AL, 17H
         OUT DX, AL       ; Set the control word, clear the error flag, full
                            duplex, DTR valid
         CALL DELAY
         MOV AX, DATA
         MOV DS, AX
         LEA BX, BUFF
         MOV CX, 10
         MOV DX, 301H     ; Read status, query RxRDY
  STATUS:
         IN AL, DX
         TEST AL, 02H     ; RxRDY = 1?
```

```
            JZ STATUS
            MOV DX, 300H     ; R_xRDY = 1, Xlsread
            IN AL, DX
            MOV [BX], AL     ; Enter and save data
            INC BX
            MOV DX, 301H
            IN AL, DX
            TEST AL, 38H     ; Is there any error in the judgement
            JNZ ERR          ; error handling; error handling
            LOOP STATUS
            JMP EXIT
    ERR: ...
    EXIT: ...
```

2. Examples of two microcomputers communicating with each other through 8251A

[**Example 7-13**] Implement the communication between two PC with short distance by 8251A. The hardware connection is as shown in Figure 7-26. Due to the short distance communication, so no Modem, two computers can be connected directly by RS-232C, and the communication of both sides as a data terminal equipment DTE. Because of RS-232C interface standard, it is necessary to add a conversion circuit. And that the other side is ready for communication, so we can not use the 4 root contact signal, only the 8251A $\overline{CTS}$ ground.

The data line $D_0 \sim D_7$ and 8251A 8086 in the figure are connected to $C/\overline{D}$ signal A_1 and 8251A 8086 processor, so the requirements of the 8251A port address for my address. The address decoder that gets the decoding signal is 200H, then 8251A control port address 202H, data port address for 200H.

Two machine A, B can be carried out between simplex or full duplex. CPU interface can transmit data according to the query or interrupt. In this case by simplex communication, query mode, asynchronous transmission. For initialization and control procedures of the sending end and receiving end.

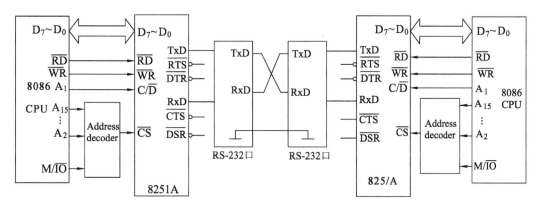

Figure 7-26 Simplified block diagram of the connection between two microcomputers for mutual communication

The initialization program consists of two parts:

(1) define one side as a transmitter. The sender's CPU outputs one byte of data to the

8251A in parallel when each query is valid for TXRDY.

(2) the receiver is defined as another side. The receiver's CPU is valid for each query to the RXRDY, and a byte data is entered from the 8251A until all data is transmitted.

The sending end initialization program and the sending control program are as follows:

```
SENT:
    MOV DX,202H         ; Set control port address
    MOV, AL, 7FH        ; define 8251A as asynchronous, 8 bit data, 1 bit
                          stop bit
    OUT, DX, AL         ; even check, take the baud rate coefficient of 64
    MOV AL, 11H         ; clears the error flag to allow sending
    OUT DX,AL
    MOV DI, sends the first address of the block of data
                        ; sets the address pointer
    MOV CX, sends the number of bytes of data blocks
                        ; sets the initial value of the counter
NEXT: MOV DX,202H
    IN AL,DX
    AND AL,01H          ; Is the query valid for TXRDY?
    JZ NEXT             ; TXRDY = 0, invalid wait
    MOV DX,200H
    MOV, AL, [DI]       ; output a byte of data to the 8251A
    OUT DX,AL
    INC DI              ; modify the address pointer
    LOOP NEXT           ; do not transfer, then proceed to the next
    HLT

; The receiver initialization program and the receive control program are
  shown below:
RECV: MOV DX,202H
    MOV, AL, 7FH        ; initializes 8251A, asynchronously, 8-bit data
    OUT, DX, AL         ; 1 bit stop bit, parity check, baud rate factor 64
    MOV AL, 14H         ; clears the error flag and allows for receive
    OUT DX,AL
    MOV DI, receive the first address of data block
                        ; and set the address pointer
    MOV CX, the number of bytes to receive data blocks
                        ; set the counter initial value
COMT: MOV, DX, 8251A control port
    IN AL,DX
    ROR AL,1            ; Is the query valid for RXRDY?
    ROR AL,1
        JNC COMT        ; invalid waiting
    ROR AL,1
    ROR AL, 1           ; further check if there is parity error when valid
    JC ERR              ; error handling when error occurs
```

```
        MOV DX,200H
    IN AL,DX                ; Without error, enter a byte to receive the data
                              block
    MOV [DI],AL
    INC DI                  ; modify the address pointer
    LOOP COMT               ; do not transfer, then proceed to the next
    HLT
    ERR:…                   ; Error handling
```

7.4 Programmable timer/counting interface chip 8253/8254

A microcomputer system is often used for timing signal. Such as timing and dynamic memory system calendar time refresh timer. In the field of automatic control systems, often requires the system at a certain time interval to the environment temperature and humidity information acquisition, or count the number of external processes. In daily life, intelligent electric rice cooker, washing machine and TV use the timer; currency-counting machine and speed printing machine also use counter. The essence of the timing or counting is the same, is to count pulse signal. The pulse timer is the internal clock signal standard, constant cycle count, determines the size of the length of time If the count object is related to the external pulse signal (the cycle can not be equal), this is the counter at this time.

Generally speaking, the implementation of timing and counting can be divided into two methods: software and hardware.

The software method is completed through delay subroutine, when the timing time is long, and need to be done by double circulation, as discuss in the sixth chapter, you can use the software to eliminate the delay jitter when you do keyboard scanning. Software timing method doesn't need too much hardware, control is simple and convenient, but in a period of time, CPU can do nothing, reduces the utilization of the computer, and the timing accuracy is low.

The hardware method is realized by non-programmable hardware timer or programmable hardware timer/counter. The non-programmable hardware timing is the original counter which connected in circuit, timing and scope can not change, is not flexible enough. The programmable hardware timer/counter can be timer/counter through instruction, the required time will automatically generate an output. CPU in the process can do other work, it improves the work efficiency of CPU with high precision. By this method of precise timing, flexible and easy to use, widely extensive application.

The Intel 8254 is a widely used in microcomputer programmable timer/counter chip. The early use of Intel 8253 as the system timer/counter, 8254 is enhanced 8253, its internal structure, working mode and methods are basically the same compare with 8253. The main difference between them is the count frequency, 8254 higher than 8253, increasing the internal readout control and status word. This section introduces the 8254 chip.

7.4.1 The internal structure of 8254

The 8254 chip is a programmable timing/counting chip is very extensive, its main function includes timing and counting. The microcomputer dynamic memory refresh circuit, system clock source technology and sound system is made up of 8254 chip to complete.

The internal structure of 8254 is shown in Figure 7-27. It mainly consists of the following main parts:

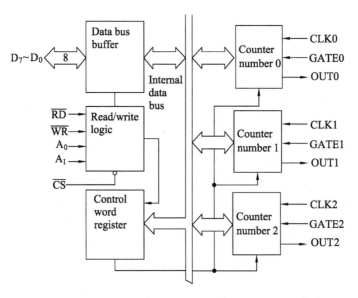

Figure 7-27 8254 internal structure

1. The data bus buffer

The 8-bit bidirectional three state buffer connected by the 8254 and the CPU data bus is used to transmit the control information of CPU to 8254, the data information and the status information of CPU read from 8254, including the real-time count value of a certain time.

2. The read/write control logic

8254 and on the internal control chip select register read/write operations, it receives an address signal sent by the CPU to achieve chip select, internal counter selection and control of read/write operations.

3. The control word register

During initialization of 8254, the write word is written by the CPU to determine how the counter works. This register can only be written and cannot be read out.

4. The counter $0^\#, 1^\#, 2^\#$

This is the three independent, counter/timer. The same structure of each counter internal structure as shown in Figure 7-28, contains a 16-bit initial count register, to store the initial count value, a 16-bit counter minus 1 and a 16 bit output latch lock output. In the process of counter latch work, follow the change count and change the count in command received by the CPU to read, numerical value to latch the meter for the CPU to read, read after the output latch and follow by 1 counter changes.

CHAPTER 7 PROGRAMMABLE INTERFACE CHIP

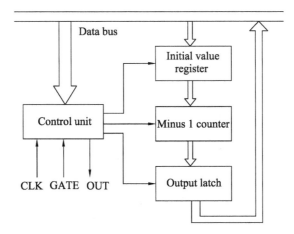

Figure 7-28 Internal structure of counter

7.4.2 8254 external pin

The 8254 chip is a dual in-line integrated circuit chip with 24 pins. The pin distribution is shown in Figure 7-29. The 24 pins of the 8254 chip are divided into two groups, one is oriented to the CPU, the other is oriented to the external device.

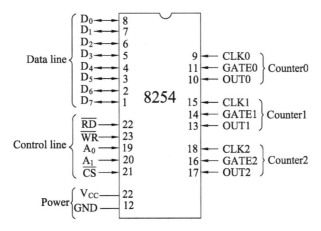

Figure 7-29 8254 pin

1. Interface signal with CPU

(1) $D_7 \sim D_0$: 8-bit bidirectional three state data pin, which is connected with the data line of the system, transmits control, data and status information.

(2) $\overline{RD}$: the read control signal input pin from the CPU is active low.

(3) $\overline{WR}$: from the CPU write control signal input pin, active low.

(4) $\overline{CS}$: the chip select signal input pin is active low.

(5) A_1, A_0: the address signal input pins, which is used to select the 8254 chip counter and controls word register. The relationship between the state of A_1, A_0 and the 8254 port address is shown in Table 7-4.

255

Table 7-4 8254 read/write operation logic table

$\overline{CS}$	$\overline{RD}$	$\overline{WR}$	A_1	A_0	Register selection and operation
0	1	0	0	0	write counter0[#]
0	1	0	0	1	write counter1[#]
0	1	0	1	0	write counter2[#]
0	1	0	1	1	Write control word register
0	0	1	0	0	Read counter 0[#]
0	0	1	0	1	Read counter 1[#]
0	0	1	1	0	Read counter 2[#]
0	0	1	1	1	no-operation
1	×	×	×	×	prohibition of use
0	1	1	×	×	no-operation

If connected with the microcomputer 8254 and 8-bit data bus, as long as the A1A0 respectively with a minimum of two $A_1 A_0$ of the address bus can be connected. For example, in the PC/XT CPU in 8088 as the upper part of the address bus ($A_9 \sim A_4$) port for I/O decoding, select the I/O chip chip select signal to form the lower part ($A_3 \sim A_0$) for addressing each chip internal port. If the base address port 8254 for 40H, 0,1,2 and counter control word register address port are respectively 40H, 41H, 42H and 43H.

V_{CC} and GND: +5V power and ground pin.

2. Interface signal to external device

CLK_i: C_i =0,1,2. The I-th counter pulse input pin, CLK can be a system clock pulse, can also be provided by other sources. If the input pulse is constant cycle clock, 8254 general work in a regular manner, if the input is pulse cycle uncertain, or only cares about the number of pulses instead of pulse interval at this time, 8254 commonly used as a counter clock signal. The clock signal frequency of 8254 is 0 ~ 12 MHz. the frequency of the 8255 clock signal is 0 ~ 2.6MHz.

$GATE_i$: (i = 0,1,2): the I-th counter gating signal input pin the counter function is related to a gating signal working mode, used to control the counter to start or stop. When the GATE is high, allowing the counter, when GATE is low, no counter work. Two or more than two counters when used, the availability of signal synchronization, can also be used to synchronize with external signals.

OUT_i: (i =0,1,2): I-th timer/counter singal output pin, the output signal of the form is determined by the counter mode, and the output signal can be used to trigger other circuit, or as an interrupt request signal sent to the CPU.

7.4.3 How does the 8254 work

8254, there are 6 modes of work, the working conditions are different in different ways, the output waveform is different, the requirements of the GATE signal are different, but the process of each method is similar:

(1) Set the counter's mode of operation. Write the control word to the counter. All the control logic circuits are reset immediately, and the output OUT is in the initial state. The initial state is not necessarily the same for different modes.

(2) Set the initial value of the count. Write the count initial value to the counter port. If the initial value is 8 bits, just write once, the initial value is 16 bits, you must write it two times.

(3) Hardware boot. GATE produces the start counter from low to high rising edge, only mode 1 and mode 5 require hardware start.

(4) When the initial value is written, the falling edge of the next CLK is counted and the initial value is reduced to 1 minus counter.

(5) Each input a clock, 1 minus counter, reduces 1 count.

(6) Counting process is over.

Usually, at each clock pulse rising edge of CLK, sampling GATE. gating signal in different ways, the gating signal effect of GATE. In general GATE =1, GATE =0, the counter, the counter prohibited. Please readers attention in every way, with the GATE signal.

The features of each approach are described below:

1. Mode 0-counting ends, interrupts

As shown in Figure 7-30 the waveform of 0, when the control word write control word register, the output OUT becomes low; when the count is written after a counter starts counting the falling edge of CLK. In the counting process, OUT remains low, when the count to 0, OUT becomes high, the signal is in conformity with the interrupt request signal, so it is called the end count level control counting process interrupt mode is GATE.

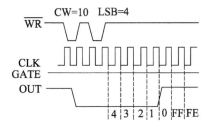

Figure 7-30 Mode 0 waveform

As can be seen in Figure 7-30, the working mode 0 has the following characteristics:

(1) The counter is only counted again. When the count is up to 0, the count is not restarted, and the OUT remains high until a new value is entered, the OUT becomes low, and a new count is started.

(2) The counter starts counting at the next CLK after the write value command. If the counter is set to N, the output OUT can be higher after N +1 pulses.

(3) In the counting process, the GATE signal can be used to control the pause. When GATE =0, the pause count is continued. When GATE =1, the count continues.

(4) The counting process can change the value, and this change is effective immediately, divided into two types: if 8-bit counts, write a new value after a falling edge of CLK according to the new count value; if 16-bit counts, then stop counting in the writing of the first byte, write. Second bytes after a falling edge of CLK according to the new value count.

2. Mode 1-programmable hardware trigger monostable negative pulse mode device

As shown in Figure 7-31 the waveform 1, CPU to 8254 OUT after writing the control word becomes high, and keep writing, the count is not immediately after the count, only when the input GATE signal after the rising edge, a falling edge of CLK began to count, low OUT, count to 0, OUT became high at this time, the end of count. Then a GATE rising edge, counter began to count, the output OUT becomes low again. Therefore, each counting period OUT to output a negative pulse.

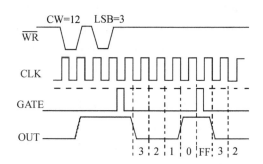

Figure 7-31 Mode 1 waveform

Mode 1 has the following characteristics:

(1) After writing the initial value, even if the GATE is 1, the counter does not work immediately. The counter on the rising edge of the GATE pin is required to count. This is called hardware start up.

(2) The negative pulse width of OUT output is counting initial value * CLK cycle.

(3) The output is controlled by gated signal GATE, in three cases:

① After counting to 0, the GATE pulse is added, then the count is restarted, and the OUT becomes low.

② In the counting process, the GATE pulse starts counting again from the falling edge of the next CLK, and the OUT remains low.

③ After changing the count value, only when the GATE pulse is started, the new value is counted, otherwise the count process is not affected and continues. That is, the change of the new value starts from the next GATE.

(4) The count value is valid many times. Every GATE pulse is automatically loaded into the numerical value and starts counting from scratch. Therefore, at the time of initialization, the count value is written once.

3. Mode 2-rate generator

As shown in Figure 7-32, in this way 2, the output of the CPU control word, the output OUT becomes high, write count after a falling edge of CLK began to count, count to 1, the output OUT becomes low, after a CLK, OUT recovery is high, to counter start counting. In this way, GATE remains high, just write a count, can continuously periodically output negative pulse.

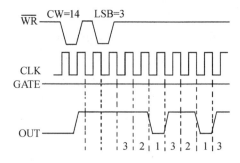

Figure 7-32 Mode 2 waveform

Mode 2 has the following characteristics:

(1) The period of the OUT signal is N CLK cycles, and the ratio of the high and the low is N-1 : 1.

(2) In counting, when the GATE is low, pause count, revert to high and count again from the initial value (Note: this mode is different from mode 0, and mode 0 is continued

counting).

(3) Change the count value in the counting process, then the new count value starts at the next CLK falling edge, counting in the same way as the new value, 1.

4. Mode 3-square wave rate generator

Mode 3 waveform shown in Figure 7-33, similar to the way 2, are N CLK cycle output continuous waveform, but the high and low level are different. CPU write control word, the output OUT higher, after writing the count value, start counting.

The initial count is even, each CLK cycle counter, 2 count, count to 0, the OUT output becomes low, the reload count value minus 2 count, when the count to 0, the output is high, then loading the initial 2 count, so the cycle continues.

The initial count is odd, the initial value reduced by 1 (even) numerical device counter, each falling edge of the CLK pulse count by 2, reduced to 0 after the next CLK, OUT goes low, then still in the initial count minus 1 values after reloading count, count to 0. OUT becomes high level, so the automatic repeat count.

Compared with the method 2, mode 3 has the following differences:

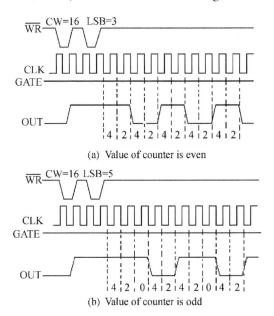

Figure 7-33 Working waveform of counter in mode 3

(1) If the numerical value is even, then output standard square wave, the high-low level is N/2 CLK cycle; if odd number, output (N+1)/2 CLK cycle high level, (N-1)/2 CLK cycle low, approximate to square wave.

(2) Change the count value does not affect the counting process of current in the counting period, under normal circumstances, a new count is at the end of the week after the half into the counter. But if the first half cycle GATE trigger signal, then the pulse after loading a new value to start counting.

5. Mode 4-Software triggered strobe signal generator

As shown in Figure 7-34, in this way 4, when the CPU control word is written, OUT immediately becomes high, write count a CLK start counting, when the count to 0, low OUT, after a CLK pulse, OUT becomes high. This count is a one-time (with 0 similarity), only when writing a new count after the start of the next count.

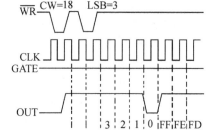

Figure 7-34 Mode 4 waveform

Mode 4 has the following characteristics:

(1) When the numerical value is N, the interval N +1 CLK pulses output a negative pulse (count once valid).

(2) GATE =0, count off, GATE =1, restore count again.

(3) The new value is reloaded in the count process, and the value is immediately valid (if the 16 bit value is counted, the first byte is stopped, and the second byte is loaded to start counting by the new value).

6. Mode 5-hardware triggered strobe signal generator

The waveform of mode 5 is shown in Figure 7-35, and the output waveform is the same as mode 4. However, the initial value is not immediately started after the initial value is written. It is only possible to count the signal at the rising edge of the GATE when the trigger signal arrives

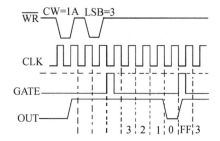

Figure 7-35 Mode 5 waveform

Compared with mode 4, mode 5 has the following characteristics:

(1) After the end of the count, if there is another GATE trigger signal, reload the count value and start counting.

(2) If another GATE pulse is added in the counting process, the initial value is reset and the output is unchanged, that is, the count value is valid several times.

(3) If the count value is modified in the counting process, the value is loaded after the next GATE pulse and is counted at this value.

7. Summary of ways of working

Although there are 6 modes of work in 8254, there are many similarities:

(1) The output waveform of the mode 2,4,5 is the same, all of which is a negative pulse with the width of one CLK cycle, but the mode 2 works continuously. The mode 4 is triggered by software, and the mode 5 is triggered by hardware.

(2) 5 and 1 are used to trigger the GATE rising edge, the working process is the same, but the output waveform, 1 output is the width of N CLK cycle active low pulse (counting process output is low, and 5) the width of the output pulse is a negative cycle CLK (the counting process output is high).

(3) The initial state of the output OUT, the mode 0 is the output after the write mode is low, and the rest of the way, after writing the control word, the output is changed too high.

(4) Either method can start counting after the initial value of the write count. The mode

0,2,3,4 starts counting after the initial value of the write count, while mode 1 and mode 5 require external trigger start before they start counting.

(5) 6 working modes, only mode 2 and mode 3 is continuous counting, the other way is one count, continue to work to restart, 0,4 by software startup (need to override the initial count), 1,5 by the hardware boot (GATE rising edge).

(6) The role of the gating signal; through the gating signal GATE, counting process can be a counter intervention in 8254. In different working modes, the gating signal plays a role in different ways. 0,2,3,4 level plays a role, 1,5 is a rising edge effect, along the way all play a role in the rise and 2,3 level (high level counting the rising edge, start counting).

7.4.4 Control word for 8254

1. Mode control word

The 8254 has a 8-bit mode control word, as shown in Figure 7-36.

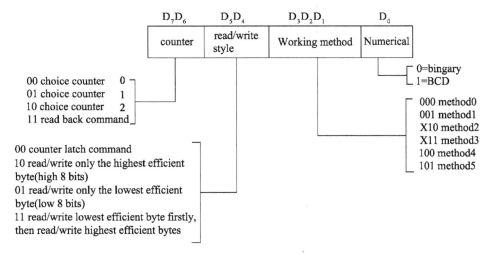

Figure 7-36 8254 control word

Among them:

D_0: select the control system. As of 1, show that the BCD code for timer/counter; otherwise, the binary timing/counting.

$D_3 \sim D_1$: work mode selection control. 000: mode 0; 001: mode 1; X10: mode 2; X11: mode 3; 100: mode 4; 101: mode 5.

$D_5 D_4$: read and write format. 00: count latch command, latch corresponding counter current count value; 01: read/write high 8 commands; 10: read/write high 8 command; 11: first read/write low 8, read again write high 8 command.

$D_7 D_6$: counter selection control. 00 counter 0; 01: 1 10: 2 counter; counter; 11: read command is back, see the specific format of readback control words, 8253 for illegal operation.

In addition, in binary, the initial value is written: 0000H ~ FFFFH, the maximum value is 0000H, the number of representatives of the 65536 count; in decimal, the initial value is 0000H to write 9999H, where 0000H is the maximum value of counting times on behalf of the 10000. to 12 of the maximum count of what is 0000H instead of FFFFH and 9999H? 8254 is because the counter, the default 0000H is 10000H, this is 10000H in binary (65536), next is

10 in decimal The minimum initial value of the 0 mode 2 and the method 3 is 2, and the minimum initial value of the other methods is 1.

[**Example 7-14**] Set 8254 port address from 40H to 43H, to make the counter 1 to work in the way 0, using only 8 binary count, counting the value of 80, initialize programming

Control word is: 01010000B = 50H.

The initialization procedure is as follows:

```
MOV AL,50H
OUT 43H,AL        ; Output control word from control port
MOV AL,80
OUT 41H,AL        ; Output count initial value from counter port
```

Note: if the BCD count is used, the control word is 51H, and the output value is 80H.

[**Example 7-15**] Set the port address of 8254: 0E0H ~ 0E3H, if you use counter 2, work in mode 2, press BCD count, count value 1050, initialize programming.

Control word is: 10110101B = 0B5H.

The initialization procedure is as follows:

```
MOV AL,0B5H
OUT 0E3H,AL       ; Set the mode of operation first
MOV AL,50H
OUT 0E2H,AL
MOV AL,10H        ; Write the initial value to counter 2. Write the lowest
                    8 bits and then write the 8-bit high
OUT 0E2H,AL
```

2. Read control word

The read control word lattice can read the internal count value or state of the counter, the read control word is only suitable for the 8254 chip, and the 8253 does not have the control word. The read control word format is shown in Figure 7-37.

D_7D_6	D_5	D_4	D_3	D_2	D_1	D_0
11	0:Latched value	0:latch mode	1: $2^\#$	1: $1^\#$	1: $0^\#$	0

Figure 7-37 8254 read control word

The control word D_7D_6 must be 1 read, D_0 must be 0, the three together constitutes a sign read control word 8254. $D_5 = 0$ latch count, so CPU $D_4 = 0$ will read; state information is latched into the state flag register; $D_3 \sim D_1$ is used to select D_5 and D_4 bit latch counter, respectively corresponding to the counter 2, the counters 1 and 0, both the latched state or latch count, do not affect the count.

The read command enables simultaneous latch of the 3 counter's value/status information, and when CPU reads the value of a counter/state value, the output register of the counter is automatically unlocked, and other counters are not affected.

Read write control word, count/state value is latched, can be output from the corresponding counter latch latches the readout. At the same time if the count/state output value, first read the counter latch, get the state value format as shown in Figure 7-38, second read the current value is latched counter.

[Example 7-16] Set 8254 port address as: 0E0H ~ 0E3H, read the 16 bit counter of counter 2 and save it in the CX register, programmed as follows:

```
MOV AL,0C8H          ; The read control word is 11001000B, and the count
                       value of the latch counter is 2
OUT, 0E3H, AL        ; latched values
IN, AL, 0E2H         ; read first, 8-bit low
MOV, CL, AL          ; keep low eight bits
IN, AL, 0E2H         ; post read high 8-bit
MOV, CH, AL          ; save eight bits, read the result in CX
```

3. Status word

The status word format is shown in Figure 7-38.

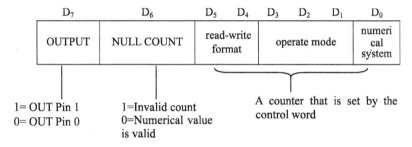

Figure 7-38 8254 status word

D_7 said the output state of the output OUT pin, D_7 =1 OUT said the end of current high level output, D_7 =0 OUT said the current low level output end. D_6 indicates whether D_6 =0 has been loaded into the initial count, said it had read into the initial value, the count value effectively, D_6 =1 said the count. D_5 ~ D_0 is invalid by the way you determine the control word, a corresponding with the same control word.

7.4.5 Application of 8254

1. 8254 application in PC

[Example 7-17] In IBM PC/XT, 8254 is used as a timing counter circuit, as shown in Figure 7-39, and its three counters act as:

The counter 0 operates in mode 3, the $GATE_0$ is fixed to a high level, and the OUT_0 is used as an interrupt request signal to the zeroth level IRQ_0. Of the 8259A interrupt controller, the timing interrupt (about 55ms) used for the time reference of the timekeeping clock.

Counter 1 in mode 2, $GATE_1$ fixed at a high level, the output of the OUT_1 through a D flip-flop as DMA controller 8237A channel 0 DMA request $DREQ_0$ for timing (about 15US) start refresh dynamic RAM, so that there can be 132 refresh in the 2ms, more than 128 times (128 times the minimum system requirements the).

Counter 2 in mode 3, $GATE_2$ 8255A PB_0, OUT_2 output square wave of 1KHz. OUT_2 and 8255A PB_1 output control signal through a control gate, amplified and sent speakers, so that the use of PB_0, PB_1 to control the long sound or tone is high at the time.

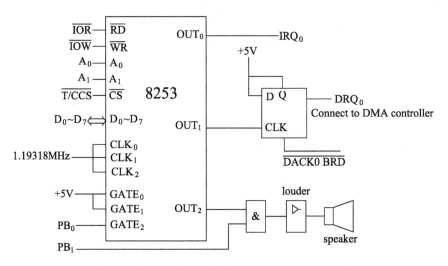

Figure 7-39 8254 application wiring diagram in PC

Suppose the counter input clock signal CLK frequency is F_{clk}, the cycle is T_{clk}, the output signal OUT frequency is F_{out}, the cycle is T_{out}, the counting initial value is N, then has:

$T_{out} = T_{clk} \times N$ standard $N = T_{out}/T_{clk} = F_{clk}/F_{out}$

The timing time of the counter 0,1,2 can be calculated according to the formula

The 8254 address is 040H ~ 043H, ROM-BIOS of 8254 programming are as follows:

(1) Counter 0 is used for timing interrupts, and initialization procedures are as follows:

```
MOV AL,00110110B      ; 00110110(binary system)
OUT 43H,AL
MOV AL,0              ; The initial value of the count is 2^16
OUT 40H,AL
OUT 40H,AL            ; timed for:840NS x 216 =55MS,Frequency is18.2Hz
```

(2) The counter 1 is used for timed DMA requests, and the initialization procedure is as follows:

```
MOV AL,01010100B      ; 01010100(binary system)
OUT 43H,AL
MOV AL,12H            ; The initial value of the count is 18D, and the
                        time is 840NS * 18
OUT 41H,AL
```

(3) The counter 2 is used to generate the square wave of the 1KHz to be transmitted to the speaker, and the sound subroutine is BEEP, as shown below:

```
; The entry parameter: the value of BL BL to control the length and length of
    sound, BL =6, long sound, BL =1, short sound.
    BEEP PROC NEAR
         MOV   AL,10110110B    ; 10110110(binary system)
         OUT   43H,AL
         MOV   AX,0533H        ; The initial value of the count is
                                 1331OUT
               42H,AL
```

```
            MOV     AL,AH
            OUT     42H,AL
            IN      AL,61H          ; Take the B port data for 8255A
            MOV     AH,AL           ; Deposit in AH
            OR      AL,03H          ; send PB0 PB1 =11
            OUT     61H,AL          ; Output to the 82255 port of the B to
                                      make the speaker sound
            SUB     CX,CX           ; cycle count
    G7:'    LOOP    G7
            MOV     BH,0
            DEC     BX              ; The delay time is controlled based on
                                      the value of BL
            JNZ     G7
            MOV     AL,AH           ; Restore the B port value of 8255A and
                                      stop sounding
            OUT     61H,AL
            RET
    BEEP    ENDP
```

2. Wave out

[Example 7-18] 8086 is CPU, through 8254 counters 0, each 2ms outputs a negative pulse. Set CLK to 2MHz, port 8254 addresses 0C0H, 0C2H, 0C4H, and 0C6H. write programs to implement this function.

Analysis: the known clock frequency $F_{clk} = 2MHz$, the output waveform of the cycle T_{out} is 2ms, then count the initial value of N is calculated as follows:

$$N = T_{out} \times F_{clk} = 2 \times 10^{-3} \times 2 \times 10^{6} = 4 \times 10^{3}$$

The subject requirements per 2ms outputs a negative pulse, mode 2 and 3 can be used; the count value is 4000, can use binary counting or BCD count, so the control word can be 00110100B, 00110110B, 00110101B and 00110111B4 of any word in a way.

An initialization code is as follows:

```
    MOV AL,34H          ; 00110100B
    OUT 0C6H,AL
    MOV AX,4000         ; The binary count initial value is 4000, and the BCD
                          count initial value is 4000H
    OUT, 0C0H, AL       ; first send low eight
    MOV AL,AH
    OUT 0C0H, AL        ; sending eight
```

Thinking: if the output signal cycle is 20ms (ie Fang Bo output 50Hz, set to work 2), CLK instead of 4MHz, software and hardware design and how?

Analysis: N = 4MHz * 20ms = 80000 exceeds the maximum initial count of 65536, must be considered by the two counter cascade, is the first level of the OUT output as second CLK input, second OUT output level for the final results. More than two level, followed by analogy. In this case only the calculated N decomposed into N1, N2,⋯(N1 * N2 *)⋯ = N) as the counting initial value at all levels.

This example can be decomposed into 4 * 2000, optional two counter cascade. The program is omitted.

Exercises 7

7.1 Try to analyze the main differences between 8255A mode 0, mode 1 and mode 2, and indicate what kind of applications they are suitable for.

7.2 When the A port of 8255A works in mode 2, what functions does port B fit in? Write the control words for various combinations at this time.

7.3 If the port A of 8255A is defined as mode 0, input is defined; port B is defined as mode 1; output is defined; the upper half of port C is defined as mode 0, and output is attempted. Write the initialization program (port address 80H~83H).

7.4 The following use of a hypothesis 8255A: port A for 0 input, port B is 0. This output connected CPU 8086 address lines A_1, A_2 are respectively connected to 8255A A_0, A_1, and the $\overline{CS}$ chip from $A_3 A_4 A_5 A_6 A_7 = 00101$, try to write the 8255A port address and initialization procedures.

7.5 What is asynchronous serial communication? What is synchronous serial communication?

7.6 Data transmission hypothesis serial asynchronous communication rate is 56000 bit/s, each frame has 7 data bits, 1 bit parity and 1 stop bit per second, it can transmit up to how many characters? Write the transmission direction of binary receiving side sends the '7' bit stream.

7.7 What registers are included within the 8251A? Examples are given to illustrate their function and how they are used.

7.8 What ports are there within the 8251A? What are their respective functions?

7.9 What are the types of 8251A pins? Their functions are described separately.

7.10 8251A is known to send data format: 7 data bits, parity, 1 stop, control the baud rate factor 64. 8251A port address is 3FAH, the data port address is 3F8H. try to write with query method and interrupt method for sending and receiving data communication program.

7.11 If the 8251A clock is clocked at 38.4KHz and its pins $\overline{RTS}$和$\overline{CTS}$ are connected, the initialization program that meets the following requirements is tried: 8251A addresses 02C0H and 02C1H.

① Simplex asynchronous communication, each character data is 7 digits, the stop bit is 1 bits, parity, baud rate is 600 bit/s, sending.

② Simplex synchronous communication, each character is 8 data bits, no parity, in synchronous mode, dual synchronous character, synchronous character of 16H, received permission.

③ Explain the application of timing and counting in actual systems. Are there any connections and differences between these two?

④ What are the implementations of timing and counting? What are the characteristics of each?

⑤ Try to explain the internal structure of the timer/counter chip Intel 8254.

7.12 What does the timer/counter chip Intel 8254 take in several port addresses? What are the ports each corresponding to?

7.13 How many jobs does the 8254 chip have? What are the characteristics of each method?

7.14 Counter 0~2 and control the port address of a system are defined for FFF0H~FFF3H. 8254 chip counter 0 in mode 2, $CLK_0 = 2MHz$, OUT_0 for the 1kHz wave output rate; the definition of the counter L in mode 0, CLK_1 external count input events, each expiration of the 100 sent to the CPU interrupt request. Try to write 8254 the counters 0 and 1 of the initial program.

7.15 Try to write a program to make the IBM PC system board audible circuit to send 200Hz to 900Hz frequency continuous change alarm sound.

7.16 8254 of the 2 work in the counter count mode, an external event introduced from CLK2, 2 per meter counter 500 pulses are sent to the CPU interrupt request, CPU in response to the interrupt after re write count, start counting, keep every 1 seconds to CPU an interrupt request. Assume the following conditions:

① 8254 counter 2 works in mode 4.
② External counting event frequency is 1kHz.
③ Interrupt type number is 54H.
④ 8254 Each port address is 200H~203H.
⑤ Interrupt management by 8259A.

Try to write programs to complete the above tasks, and draw a hardware connection diagram.

REFERENCES

［1］杨文显. 现代微型计算机原理与接口技术教程［M］. 2 版. 北京：清华大学出版社，2012.

［2］钱晓捷. 16/32 位微机原理、汇编语言及接口技术教程［M］. 北京：机械工业出版社，2012.

［3］马维华. 微机原理与接口技术［M］. 3 版. 北京：科学出版社，2016.

［4］杨全胜，胡友彬. 现代微机原理与接口技术［M］. 3 版. 北京：电子工业出版社，2013.

［5］杨季文，等. 80x86 汇编语言程序设计教程［M］. 北京：清华大学出版社，1999.

［6］朱兵，彭宣戈. 汇编语言程序设计图文教程［M］. 北京：北京航空航天大学出版社，2009.

［7］赵志诚，段中兴. 微机原理及接口［M］. 北京：北京大学出版社，中国林业出版社，2006.

［8］沈美明，温冬婵. IBM-PC 汇编语言程序设计［M］. 北京：清华大学出版社，1991.

［9］彭虎，周佩玲，傅忠谦. 微机原理与接口技术［M］. 2 版. 北京：电子工业出版社，2008.

［10］赵雁南，温冬婵，杨泽红. 微型计算机系统与接口［M］. 北京：清华大学出版社，2005.

［11］周明德，等. 80x86 的结构与汇编语言程序设计［M］. 北京：清华大学出版社，1993.

［12］戴梅萼，史嘉权. 微型计算机技术及应用［M］. 4 版. 北京：清华大学出版社，2008.

［13］肖洪兵. 微机原理及接口技术［M］. 北京：北京大学出版社，2010.

［14］冯博琴，吴宁. 微型计算机原理与接口技术［M］. 3 版. 北京：清华大学出版社，2011.

［15］林志贵. 微型计算机原理及接口技术［M］. 北京：机械工业出版社，2010.

［16］牟琦，聂建萍. 微机原理与接口技术［M］. 北京：清华大学出版社，2007.

［17］王庆利. 微型计算机原理及应用［M］. 西安：西安电子科技大学出版社，2009.

［18］杨素行. 微型计算机系统原理及应用［M］. 北京：清华大学出版社，2009.

［19］吴秀清，周荷琴. 微型计算机原理与接口技术［M］. 5 版. 北京：国科学技术出版社，2016.

［20］史新福，冯萍. 32 位微型计算机原理、接口技术及其应用［M］. 2 版. 北京：清华大学出版社. 2007.

［21］胡敏，张永. 微型计算机及其接口技术［M］. 北京：中国水利水电出版社，2010.

［22］龚尚福. 微机原理与接口技术［M］. 2 版. 西安：西安电子科技大学出版社，

2008.

[23] Muhammad Ali Mazidi, Janice Gillispie Mazidi, Ganny Causey. The x86 PC Assembly Language, Design, and Interfacing [M]. Fifth Edition. Publishing House of Electronics Industry, 2010.

[24] Randall Hyde. The Art of Assembly Language [M]. Published by No Starch Press, 2003.

[25] William B. Jones, Scott Jones. Assembly Language for the IBM PC Family [M]. 2nd Edition. Published by Scott Jones, 1997.

[26] Barry B. Brey. The Intel Microprocessor 8086/8088, 80186/80188, 80286, 80386, 80486, Pentium, Pentium Pro Processor, Pentium II, Pentium III, Pentium 4 Architecture, Programming, and Interfacing [M]. Seven Edition. Published by Prentice Hall, 2006.

APPENDIX

ASCII Value	Control Character	ASCII Value	Control Character	ASCII Value	Control Character	ASCII Value	Control Character
00H	NUT	20H	(space)	40H	@	60H	`
01H	SOH	21H	!	41H	A	61H	a
2H	STX	22H	"	42H	B	62H	b
3H	ETX	23H	#	43H	C	63H	c
4H	EOT	24H	$	44H	D	64H	d
5H	ENQ	25H	%	45H	E	65H	e
6H	ACK	26H	&	46H	F	66H	f
7H	BEL	27H	,	47H	G	67H	g
8H	BS	28H	(	48H	H	68H	h
9H	HT	29H	)	49H	I	69H	i
0AH	LF	2AH	*	4AH	J	6AH	j
0BH	VT	2BH	+	4BH	K	6BH	k
0CH	FF	2CH	,	4CH	L	6CH	l
0DH	CR	2DH	-	4DH	M	6DH	m
0EH	SO	2EH	.	4EH	N	6EH	n
0FH	SI	2FH	/	4FH	O	6FH	o
10H	DLE	30H	0	50H	P	70H	p
11H	DCI	31H	1	51H	Q	71H	q
12H	DC2	32H	2	52H	R	72H	r
13H	DC3	33H	3	53H	X	73H	s
14H	DC4	34H	4	54H	T	74H	t
15H	NAK	35H	5	55H	U	75H	u
16H	SYN	36H	6	56H	V	76H	v
17H	TB	37H	7	57H	W	77H	w
18H	CAN	38H	8	58H	X	78H	x
19H	EM	39H	9	59H	Y	79H	y
1AH	SUB	3AH	:	5AH	Z	7AH	z
1BH	ESC	3BH	;	5BH	[	7BH	{
1CH	FS	3CH	<	5CH	\	7CH		
1DH	GS	3DH	=	5DH	]	7DH	}
1EH	RS	3EH	>	5EH	^	7EH	~
1FH	US	3FH	?	5FH	_	7FH	DEL

APPENDIX

The meaning of the control characters is as follows:

NUL	VT VERTICAL TAB	SYN
SOH STARTING-OF-HEADING	FF CARRIAGE CONTROL	ETB END OF TRANSMISSION BLOCK
STX STARTING-OF-TEXT	CR CARRIAGE RETURN	CAN CANCELL
ETX ENDING-OF-TEXT	SO SHIFT OUTPUT	EM PAPER OUT
EOY ENDING-OF-TRANSFER	SI SHIFT INPUT	SUB DISPLACEMENT
ENQ ENQUIRY	DLE LEAVE A BLANK SPACE	ESC ESCAPE
ACK ACKNOWLEDGE	DC1 DEVICE CPNTROL1	FS TEXT SEPARATOR
BEL BELERT	DC2 DEVICE CPNTROL 2	GS GROUP SEPARATOR
BS BACKSPACE	DC3 DEVICE CPNTROL 3	RS RECORD SEPARATOR
HT HORIZONTAL TAB	DC4 DEVICE CPNTROL 4	US UNIT SEPARATOR
LF LINE FEED	NAK NOT ACKNOWLEDGE	DEL DELETE